Data Mining Applications Using Artificial Adaptive Systems

William J. Tastle

Editor

Data Mining Applications Using Artificial Adaptive Systems

 Springer

Editor
William J. Tastle
Ithaca College
Ithaca, NY, USA

ISBN 978-1-4939-4445-3   ISBN 978-1-4614-4223-3 (eBook)
DOI 10.1007/978-1-4614-4223-3
Springer New York Heidelberg Dordrecht London

Printed on acid-free paper

Springer is part of Springer Science+Business Media (www.springer.com)

Preface

When initially considering the content of this preface, I wanted to provide a summarization of the number and kinds of research centers in the world and make a comparison with Semeion. I was, to say the least, too conscientious in my plan. Finding the approximate number of research centers in the United States was not much of a problem (about 366) and was easily done using an Internet search. However, as I proceeded to search the available data in other countries, I quickly discovered that the task would be far more daunting than I had time available, but one item was particularly clear: there are many, many research centers across the world of varying sizes doing research in any and every field imaginable. I do not doubt that each center regularly makes a contribution to the knowledge base of humanity, and I am equally convinced that those contributions can become much too easily lost in the ether of digitalization and massive quantification of information that continues to grow at an ever increasing rate. However, there is one research center that is doing outstanding work in the field of artificial intelligence and it is to that institute that this book is directed.

The Semeion Research Centre of Rome, Italy, has been in operation since 1991 and was granted legal status recognized by the Italian Ministry for Education University and Research. It also receives financial assistance from the government in addition to grants and contracts from assorted organizations and governments. The center has a full-time staff and an international group of researchers and scholars directly associated with the organization. Some have been granted the title of "Fellow" in recognition of their accomplishments

The word "semeion" speaks well for this organization for its root is from Greek and means, putting it into proper context, *from a small quantity of data can be extracted a substantial mass of knowledge given the presence of prepared minds and an innovative spirit for discovery*

Semeion is directly involved in a series of research initiatives:

- Basic research oriented to the conception and design of artificial organisms representing adaptive systems based on Artificial Neural Networks and evolutionary algorithms for the simulation, prediction, and control of processes and phenomena

- Applied research with a focus on the construction and application of intelligent computational models in the biomedical, financial, and social fields
- Education of researchers on the methodologies and techniques of the application of Artificial Adaptive Systems to different research fields
- The distribution of research models, software, projects, and scientific testing invented inside Semeion
- Publication of scientific discoveries based on the results of research endeavors and successful experimentation carried out by Semeion's researchers, both nationally and internationally

The motivation for this book came from a conference of the North American Fuzzy Information Processing Society (NAFIPS 2010) in Toronto, Canada, during the summer of 2010. Several papers dealing with issues involved with complex problem solving and very innovative methods were reviewed by the conference publication committee and it was quickly determined that the content was exceptional, certainly more than worthy of a conference presentation. The director of Semeion, Prof. Dr. Massimo Buscema, was asked to consider the publication of the papers as part of a special issue of that Society's journal. Unfortunately, the journal officials were limited to papers whose content specialized in "fuzzy set theory," and the content of these papers was somewhat peripheral to this limitation but highly focused on the area of artificial neural networks. In retrospect, this was very good for it gave Semeion researchers an opportunity to investigate other available avenues; a proposal to Springer Science underwent peer review and was enthusiastically accepted. This also gave Semeion an opportunity to publish some very recent breakthroughs in adaptive neural network technology and applications of the technology in several disciplines, particularly the medical field.

The content presented in this book is representative of the exceptional work accomplished by Semeion researchers and is also a means by which that organization can make others more informed of the opportunities available through collaborative ventures with other individuals and research institutes.

New York, USA   William J. Tastle

Contents

Chapter 1
Assessing Post-Radiotherapy Treatment Involving Brain Volume Differences in Children: An Application of Adaptive Systems Methodology

Massimo Buscema, Francis Newman, Giulia Massini, Enzo Grossi, William J. Tastle, and Arthur K. Liu

1.1 Introduction

Perhaps the most unwelcome news one can hear from one's physician is that of the identification of a tumor and it is arguably far more painful to a parent when the news affects a young child. One standard method of treatment involves the application of radiation to the brain in an effort to shrink or otherwise eliminate the tumor. Diseased cells are destroyed in this manner, but it is well known that healthy brain cells are also destroyed, though at a lesser rate.

Research suggests that many children treated with Cranial Radiotherapy experience cognitive, educational and behavioral difficulties. The relation between changes in volume of specific brain regions after radiotherapy and the degree of decline in cognitive functions, as measured with IQ is not clear, due to high variability of response and underlying non-linearity.

M. Buscema (✉)
Semeion Research Center of Sciences of Communication, Via Sersale 117, Rome, Italy

Department of Mathematical and Statistical Sciences, CCMB, University of Colorado, Denver, Colorado, USA
e-mail: m.buscema@semeion.it

F. Newman
University of Colorado, Denver, CO, USA

G. Massini
Semeion Research Centre, Via Sersale n. 117, 00128 Rome, Italy

E. Grossi
Bracco SpA Medical Department, San Donato Milanese, Italy

W.J. Tastle
Ithaca College, New York, USA

A.K. Liu
University of Colorado, Denver, CO, USA

W.J. Tastle (ed.), *Data Mining Applications Using Artificial Adaptive Systems*,
DOI 10.1007/978-1-4614-4223-3_1, © Springer Science+Business Media New York 2013

Numerous groups have used MRI to study children treated with radiation to look for brain abnormalities, although the precise mechanism of brain injury in children resulting from radiotherapy remains poorly understood. The imaging abnormalities described have included white matter changes, cortical thinning, calcifications, hemorrhagic radiation vasculopathy, moya-moya disease, and tumors (Hertzberg et al. 1997; Harila-Saari et al. 1998; Laitt et al. 1995; Paakko et al. 1994; Poussaint et al. 1995; Liu et al. 2007; Khong et al. 2006; Leung et al. 2004; Nagel et al. 2004; Ullrich et al. 2007; Kikuchi et al. 2007; Ishikawa et al. 2006; Reddick et al. 2003, 2005, 2006; Mulhern et al. 1999). Some groups have been able to correlate the imaging abnormalities with neuropsychological deficits (Reddick et al. 2003, 2005, 2006; Mulhern et al. 1999; Paakko et al. 2000). However, other studies have been unable to find such a relationship (Harila-Saari et al. 1998; Paakko et al. 1994). Possible causes for these lack of correlations or findings is that if there is an effect on cerebral anatomy, the effect is subtle or the effect is spatially localized. Small changes may be difficult to detect with review of conventional imaging by radiologists, while localized changes may be missed if the entire brain is not closely evaluated.

Newer analysis tools may allow for more sophisticated analysis of structural changes. In this work, we utilize an automated image analysis tool (Freesurfer, a freeware application offered by the Athinoula A. Martinos Center for Biomedical Imaging) that provides accurate quantitative measurements of various brain structures based on standard clinical MRI. For example, the image analysis software enables us to track post-radiotherapy the change in volumes of cerebral cortex, amygdala, hippocampus and other structures of interest. The structural volume changes are then used as input into a novel neural network algorithm to uncover which structures are the best predictors of IQ test results.

It is the purpose of this paper to analyze data acquired from 58 children who have undergone radiotherapy treatment due to the presence of a brain tumor with the goal of identifying which brain parts are more, or less, affected.

1.2   Variables Description and Methods

The dataset used in this analysis is composed of 58 young subjects (mean age 10.13 ± 5.03 years) affected by brain tumors of different origin (Table 1.1) who underwent radiotherapy sessions.

Pre-treatment and post-treatment MRI scans were automatically segmented using the Freesurfer tools (Dale and Sereno 1993; Dale et al. 1999; Fischl et al. 2002, 2004; Segonne et al. 2004). In brief, non-brain tissue is removed and the remaining brain is registered to the Taliraich atlas and volumetric segmentation of the brain is performed. The structures segmented separately for each hemisphere and include white matter, cortex, thalamus, caudate, putamen, pallidum, hippocampus and amygdale.

Differences in the volume of 18 brain segments, measured through volumetric magnetic resonance, are considered both pre- and post-treatment. The standard of success is assumed to be the individual child's post-treatment IQ. Based on the post-treatment analysis of the data it is determined that 30 subjects were measured

Table 1.1 Distribution of brain tumors in the study population

Type of tumor	No. subjects
Medulloblastoma	14
Craniopharyngioma	9
Ependymoma	7
Nongerminomatous germ-cell tumor	6
Germinoma	6
Anaplastic astrocytoma	3
Other	13
Total	**58**

Table 1.2 The average change in brain volume, by segment, after treatment

Brain segment	Average V_1 (IQ < 94)	Average V_2 (IQ ≥ 94)
Left-Cerebral-White-Matter	−0.00472830	−0.00066057
Left-Cerebral-Cortex	0.00508573	−0.00034732
Left-Thalamus	−0.00416413	0.00046304
Left-Caudate	−0.00125607	0.00555714
Left-Putamen	0.00976560	0.00424279
Left-Pallidum	−0.00335303	0.00191857
Left-Globus Pallidus	0.00156500	0.00111636
Left-Hippocampus	0.00079647	−0.00008032
Left-Amygdala	−0.00187227	0.00064261
Right-Cerebral-White-Matter	−0.00521657	0.00039893
Right-Cerebral-Cortex	0.00580363	−0.00020357
Right-Thalamus	−0.00219633	0.00170196
Right-Caudate	−0.00045183	0.00105939
Right-Putamen	0.00029533	0.00445014
Right-Pallidum	−0.00224167	0.00130579
Right-Globus Pallidus	−0.00116657	0.00176311
Right-Hippocampus	0.00472197	0.00034175
Right-Amygdala	−0.00028133	0.00020061

Each hemisphere of the brain is composed of nine segments or parts, and each is designated as being located in either the left or right hemisphere

to have an IQ of less than 94 (subsample V_1), and 28 subjects possessed an IQ equal to or greater than 94 (subsample V_2).

The relation between the age of the subjects and the post radiotherapy IQ was very low (r = −0.27). The problem is to establish the relations between brain segments volume changes and the IQ (Table 1.2).

From the table we can see that after treatment the total volume for V_1 (0.033) is much smaller than the total volume for V_2 (0.668), but there are some exceptions. The volumes associated with the right and left cerebral cortex are much larger in V_1, the volumes for the right and left hippocampus are larger in V_1 and the volumes of the right and left white matter are much smaller in V_1.

Table 1.3 The standard deviation in brain volume, by segment, after treatment

Brain segment	Std dev V_1 (IQ < 94)	Std dev V_2 (IQ ≥ 94)
Left-Cerebral-White-Matter	0.0153	0.0072
Left-Cerebral-Cortex	0.0158	0.0049
Left-Thalamus	0.0237	0.0069
Left-Caudate	0.0140	0.0161
Left-Putamen	0.0444	0.0183
Left-Pallidum	0.0370	0.0184
Left-Globus Pallidus	0.0260	0.0081
Left-Hippocampus	0.0094	0.0036
Left-Amygdala	0.0170	0.0059
Right-Cerebral-White-Matter	0.0200	0.0078
Right-Cerebral-Cortex	0.0250	0.0032
Right-Thalamus	0.0187	0.0074
Right-Caudate	0.0101	0.0071
Right-Putamen	0.0324	0.0146
Right-Pallidum	0.0271	0.0125
Right-Globus Pallidus	0.0289	0.0075
Right-Hippocampus	0.0217	0.0023
Right-Amygdala	0.0114	0.0037

1.2.1 V_1 and V_2 as Two Separate Classes

Table 1.3 shows the standard deviations associated with each brain segment. Inspection of Table 1.3 shows a variance in V_1 that is generally larger than that of V_2. One possible interpretation of this statistic is that V_1 subjects may be more difficult to predict.

Let us take all records in V_1 for which all individuals have an IQ measured at <94 and assign them to be members of Class 1, and all records in V_2 for which individuals have an IQ measured at ≥94 to be members of Class 2. The addition of these two classes constitutes dependent variables for which the square of the correlation is calculated. Table 1.4 shows the square of the correlation.

The square correlation between each variable and the classes for the two samples together is very poor. This suggests that there is no possible way to linearly classify V_1 or V_2 subjects using the 18 variables. In short, this method of analysis that relies on the use of traditional statistics leads us to the conclusion that little practical knowledge can be learned from this data. We thus turn to a set of tools developed by Semeion (Buscema 2000a, b, 2007a, 2008a, b, 2009c; Massini 2007) in an attempt to extract something useful from this otherwise marginal set of data.

1.2.2 Linear Correlation

Microsoft Excel is used to calculate the correlation between all pairs of variables; the negative correlations are highlighted

Table 1.4 Calculation of the square of the correlations

Square of the correlation	R_2 target
Left-Cerebral-White-Matter	0.0283
Left-Cerebral-Cortex	0.0515
Left-Thalamus	0.0173
Left-Caudate	0.0504
Left-Putamen	0.0066
Left-Pallidum	0.0082
Left-Globus Pallidus	0.0001
Left-Hippocampus	0.0038
Left-Amygdala	0.0097
Right-Cerebral-White-Matter	0.0334
Right-Cerebral-Cortex	0.0277
Right-Thalamus	0.0186
Right-Caudate	0.0077
Right-Putamen	0.0069
Right-Pallidum	0.0071
Right-Globus Pallidus	0.0048
Right-Hippocampus	0.0197
Right-Amygdala	0.0008
Sum	0.3026

An examination of the linear correlation shows (see page 6):

- In subsample V_1 only the cerebral cortex and hippocampus possess a positive correlation with each other and a negative correlation with the other seven segments of the brain. From this it may inferred that these two parts of the brain were modified during the radiotherapy treatment in a different and more pronounced way.
- Subsample V_2 possesses various positive and negative correlations that are clearly more distributed; that could mean that each part of the brain was less modified by the radiotherapy treatment;
- In comparing the row summations in V_1 and V_2 it is apparent that subsample V_2 has a stronger relationship in terms of covariance among the nine brain segment volumes.

These results suggest a suspicion as to the critical role radiotherapy treatment may have on the modification of the cerebral cortex and hippocampus among the subjects of the V_1 subsample.

1.2.3   Classification of the Two Classes Through Artificial Adaptive Systems

Artificial adaptive systems (AAS) utilize highly nonlinear functions in computationally expensive ways to identify relationships among variables. This technique permits us to classify V_1 and V_2 in a "blind" way using a Training and Testing

V₁

	Cerebral-White-Matter	Cerebral-Cortex	Thalamus-Proper	Caudate	Putamen	Pallidum	Globus Pallidus	Hippocampus	Amygdala	Row Sum	Variance
Cerebral-White-Matter	1.00	-0.96	0.68	0.46	0.27	0.63	0.54	-0.71	0.84	2.75	0.4659
Cerebral-Cortex	-0.96	1.00	-0.52	-0.32	-0.21	-0.53	-0.44	0.78	-0.73	-1.93	0.4433
Thalamus-Proper	0.68	-0.52	1.00	0.57	0.67	0.86	0.87	-0.37	0.78	4.54	0.3070
Caudate	0.46	-0.32	0.57	1.00	0.23	0.52	0.41	-0.27	0.35	2.95	0.1697
Putamen	0.27	-0.21	0.67	0.23	1.00	0.62	0.93	-0.18	0.35	3.67	0.1891
Pallidum	0.63	-0.53	0.86	0.52	0.62	1.00	0.83	-0.41	0.55	4.08	0.2981
Globus Pallidus	0.54	-0.44	0.87	0.41	0.93	0.83	1.00	-0.34	0.59	4.39	0.2867
Hippocampus	-0.71	0.78	-0.37	-0.27	-0.18	-0.41	-0.34	1.00	-0.40	-0.89	0.3383
Amygdala	0.84	-0.73	0.78	0.35	0.35	0.55	0.59	-0.40	1.00	3.34	0.3346
Negative correlations	2	7	2	2	2	2	2	7	2	28	2.8326

V₂

	Cerebral-White-Matter	Cerebral-Cortex	Thalamus-Proper	Caudate	Putamen	Pallidum	GlobusPallidus	Hippocampus	Amygdala	Row Sum	Variance
Cerebral-White-	1.00	-0.63	0.85	0.08	-0.75	0.91	-0.64	0.33	-0.43	0.72	0.5189
Cerebral-Cortex	-0.63	1.00	-0.63	-0.09	0.43	-0.70	0.27	0.34	0.30	0.29	0.3436
Thalamus-Proper	0.85	-0.63	1.00	0.19	-0.54	0.82	-0.36	0.26	-0.36	1.23	0.4097
Caudate	0.08	-0.09	0.19	1.00	-0.15	0.17	-0.14	0.06	-0.43	0.69	0.1569
Putamen	-0.75	0.43	-0.54	-0.15	1.00	-0.69	0.94	-0.47	0.20	-0.03	0.4612
Pallidum	0.91	-0.70	0.82	0.17	-0.69	1.00	-0.53	0.21	-0.28	0.90	0.4763
Globus Pallidus	-0.64	0.27	-0.36	-0.14	0.94	-0.53	1.00	-0.44	0.26	0.38	0.3815
Hippocampus	0.33	0.34	0.26	0.06	-0.47	0.21	-0.44	1.00	0.08	1.37	0.1934
Amygdala	-0.43	0.30	-0.36	-0.43	0.20	-0.28	0.26	0.08	1.00	0.34	0.2204
Negative correlations	4	4	4	4	4	4	4	2	4	34	3.1616

Reverse Validation Protocol TTRVP (Buscema et al. 2005). Note that the analysis is "blind" because no preparation of the data is undertaken to provide a separation of individual datum into their respective classes.

The following artificial adaptive systems were used in this experiment:

- Advanced Back Propagation – FF_BP (Rumelhart and McClelland 1986; Buscema 2008a, b; Chauvin and Rumelhart 1995);
- Adaptive Vector Quantization 1 – AVQ1_G (Buscema 2009b; Kohonen 1995; Kosko 1992);
- Sine Net –SN (Buscema et al. 2006a, b);
- Guacamole (Buscema 2008b);

  Algorithms:

- NRC (Buscema 1998b; Diappi et al. 2004);
- AutoCM, AutoBP, SCM (Buscema 2007b; Buscema and Grossi 2008; Buscema et al. 2008);
- Majority Vote (Day 1988; Buscema 2007a; Kuncheva 2004);
- Meta-Fuzzy (Buscema 1998a, 2007a);

Four types of experiments were carried out on the sample data (note that reference to the "18" means the set of brain segments, separated by hemisphere, as listed in Tables 1.1, 1.2, and 1.3):

Experiment 1: The 18 real values: The test was conducted using the real values of difference of volume among 18 pre- and post-treatment segments of the brain;

  The purpose is to test the goodness of the raw data;

Experiment 2: The 18 binary values: The second experiment only considered the sign difference in each of the 18 sections (-1, $+1$); to eliminate all possibilities of noise in the data stream, we eliminate all but the sign difference between the pre-and post-treatment volumes of the brain;

Experiment 3: The nine summations of the real values: The third experiment composes a new input vector for each subject made up by summing the real values of the same left and right part of the volume's difference. Thus, the left cerebral white matter variable was added to the right cerebral white matter variable to produce a single white matter variable. In this way, each input vector was compacted to nine components and no distinction between left and right is recognized; the purpose of this experiment is to verify if the key information contained in this very small dataset is not simply in the volume differences but in the global compensation between each right and left part of the brain;

Experiment 4: The nine modules as the summation of the real values: The fourth experiment was similar to the previous experiment, but the module of the sum is considered. The intent is to test if the sign of the value is relevant or not to the classification of brain quality.

1.2.4   Prototype Discovery Through a New Adaptive System: ACS

There exists a pressing need to develop an algorithm to clearly delineate the effects of radiotherapy treatment on the various brain segments. Such an algorithm would produce prototypes based on the V_1 and V_2 subsamples. We seek to establish suitable prototypes below.

A new Artificial Adaptive System has been developed that is able to discover the prototypes embedded in the two subsamples. The name of this system is the Activation and Competition System (ACS) (Buscema 2009a, c) and it is composed to two parts:

1. The algorithms are able to calculate the basic association among all the variables in the dataset; in the case of this particular dataset, the basic association is on the nine parts of the brain volume differences.
2. Utilizing the correlations among the variables as system constraints, dynamically generate the prototype that is embedded in the dataset.

The dataset to be processed will include the nine parts of the brain volumes differences contained in the two subsamples, along with the tag variables ("1 0" for V_1 and "0 1" for V_2), to distinguish the subsamples to which each record belongs.

1.3   The Theory of Activation and Competition System

ACS is an artificial adaptive system designed by Massimo Buscema in 2009 at Semeion Research Center in Rome (Buscema 2009a, c). It is a dynamic neural network able to merge many auto associative connection matrices that are generated by different algorithms, thus able to simultaneously consider many different types of mathematical associations that exist among the same set of variables. The results from ACS are detailed and robust.

As is characteristic of neural networks, ACS has an initial learning phase that is based on the variables under study. These variables are called units. Each unit evolves toward a new equilibrium state, called an attractor, using the vector of the connection matrices as a set of constraints.

1.3.1   Application of ACS

In this application ACS uses two different algorithms to generate its vector of connections matrices:

- The Linear Correlation Matrix:

$$W^{[L]}_{i,j} = \frac{\sum\limits_{k=1}^{N} (x_{i,k} - \bar{x}_i) \cdot (x_{j,k} - \bar{x}_j)}{\sqrt{\sum\limits_{k=1}^{N} (x_{i,k} - \bar{x}_i)^2 \cdot \sum\limits_{k=1}^{N} (x_{j,k} - \bar{x}_j)^2}}; \quad (1.1)$$

$$-1 \leq W^{[L]}_{i,j} \leq 1; \quad i,j \in [1,2,\ldots,M]$$

- The Prior Probability Algorithm:

$$W^{[P]}_{i,j} = -\ln \frac{\frac{1}{N^2} \cdot \sum\limits_{k=1}^{N} x_{i,k} \cdot (1 - x_{j,k}) \cdot \sum\limits_{k=1}^{N} (1 - x_{i,k}) \cdot x_{j,k}}{\frac{1}{N^2} \cdot \sum\limits_{k=1}^{N} x_{i,k} \cdot x_{j,k} \cdot \sum\limits_{k=1}^{N} (1 - x_{i,k}) \cdot (1 - x_{j,k})}$$

$$-\infty \leq W^{[P]}_{i,j} \leq +\infty; \quad x \in [0,1]; \quad i,j \in [1,2,\ldots,M]$$

All weight matrices coming from the various algorithms are linearly scaled between 0 and 1.

$$Ecc_i = \alpha \cdot \sum_k^{Q} \frac{\sum\limits_{j}^{M} u^{[n]}_j \cdot W^k_{i,j}}{N^{[E]}_{k,i}} \quad W^k_{i,j} > 0;$$

$$Ini_i = \alpha \cdot \sum_k^{Q} \frac{\sum\limits_{j}^{M} u^{[n]}_j \cdot W^k_{i,j}}{N^{[i]}_{k,i}} \quad W^k_{i,j} < 0;$$

$$E_i = Ecc_i + \beta \cdot Input_i \quad Input_i > 0;$$

$$I_i = Ini_i + \beta \cdot Input_i \quad Input_i < 0;$$

$$Net_i = \left(Max - u^{[n]}_i\right) \cdot E_i + \left(u^{[n]}_i - Min\right) \cdot I_i - Dec^{[n]}_i \cdot \left(u^{[n]}_i - Rest\right);$$

$$\delta_i = Net_i \cdot \left(1.0 - u^{[n]}_i \cdot u^{[n]}_i\right);$$

$$H^{[n]} = \sum_i^{M} \delta_i^2;$$

$$\begin{cases} u^{[n+1]}_i = u^{[n]}_i + \delta_i; & -1 < u^{[n]}_i < +1 \\ u^{[n]}_i > Max & u^{[n]}_i = Max \\ u^{[n]}_i < Min & u^{[n]}_i = Min \end{cases}$$

$$Dec^{[n+1]}_i = Dec^{[n]}_i \cdot e^{-\left(u^{[n]}_i \cdot u^{[n]}_i\right)}.$$

(The description of the notation used in these equations is in Table 1.5).

Table 1.5 Explanation of symbols used in equations

Symbol	Meaning
M	Number of variables – Units
Q	Number of weights in matrices
i, j, k	$i, j \in M; k \in Q$
$W_{i,j}^k$	Value of the connection between the ith and the jth units of the kth matrix
Ecc_i	Global excitation to the ith unit coming from the other units
Ini_i	Global inhibition to the ith unit coming from the other units
E_i	Final global excitation to the ith unit
I_i	Final global inhibition to the ith unit
[n]	Cycle of the iteration
$u_i^{[n]}$	State of the ith unit at cycle n
$H^{[n]}$	Number of units updating at cycle n
δ_i	Delta update of the ith unit
Net_i	Net input of the ith unit
$Input_i$	Value of the ith external input: $-1 \leq Input_i \leq +1$
$N_{k,i}^{[E]}$	Number of positive weights of the kth matrix to the ith unit
$N_{k,i}^{[I]}$	Number of negative weights of the kth matrix to the ith unit
Max	Maximum activation: $Max = 1.0$
Min	Minimum activation: $Min = -1.0$
Rest	Rest value: $Rest = -0.1$
$Decay_i^{[n]}$	Decay of activation of the ith unit at cycle n: $Decay_i^{[n=0]} = 0$
α	Scalar for the E_i and I_i, net input to each unit: $\alpha = 1/M$
β	Scalar for the external input: $\beta = 1/M$
$\int$	A small positive quantity close to zero

The minimization of H[n] is the cost function of ACS. Consequently, when H[n] < ε, the algorithm terminates.

1.3.2 Some Considerations

ACS is an artificial neural network (ANN) endowed with an uncommon architecture. Every pair of nodes is not linked by a single value but rather by a vector of weights in which each vector component is derived from a specific metric. Such a diversity of combinations of metrics can provide interesting results when each metric describes different and consistent details about the same dataset. For this particular application, the ACS is an appropriate algorithm that forces all the variables to compete among themselves in various ways.

The ACS algorithm possesses interesting properties such as being based on the weighs matrices of other algorithms. ACS uses these matrices as a complex set of multiple constraints to update its units in response to any input perturbation. Consequently, ACS works as a dynamic nonlinear associative memory. Whenever any input is set on, ACS will activate all its units in a dynamic, competitive and cooperative process at the same time. This process will end when the evolutionary negotiation among all the units finds its natural attractor.

1.3.3   Characteristics of ACS

The ACS ANN is a complex kind of CAM system (Content Addressable Memory). When it is compared to the classic associative memory systems (Rumelhart et al. 1986; Hopfield 1982, 1984; Hinton and Anderson 1981; McClelland and Rumelhart 1988; Grossberg 1976, 1978, 1981), ACS presents some new features. First, ACS simultaneously works with many weight matrices that come from different algorithms; Grossberg's Interaction and Activation Competition network (IAC) uses only one weight matrix. The ACS weight matrices represent different mappings of the same dataset and all the units (variables) are processed in the same manner; Grossberg's IAC only works when the dataset presents a specific kind of architecture. The ACS algorithm can use any combination of weight matrices coming from any kind of algorithm as long as the values of the weights are linearly scaled into the same range, typically between −1 and +1; Grossberg's IAC can work only with static excitation and inhibitions. Finally, each ACS unit tries to learn its specific value of decay during its interaction with the other units; Grossberg's IAC works with a static decay parameter for all the variables. In short, the ACS architecture is a circuit with symmetric weights (vectors of symmetric weights), able to manage a dataset with any kind of variables (Boolean, categorical, continuous, etc.), while Grossberg' IAC can work only with specific types of variables.

1.4   Discovering Hidden Links with a New Adaptive System: Auto-CM

The Auto Contractive Map (AutoCM for short) is a new Artificial Neural Network designed by Massimo Buscema in 1998 at Semeion Research Center (Buscema 2007b). The Auto-CM system finds, by means of a specific learning algorithm, a square matrix of weighted connections among the variables of any dataset.

This matrix of connections presents many suitable features:

(a) Nonlinear associations among variables are preserved;
(b) Connections schemes among clusters of variables is captured, and
(c) Complex similarities among variables became evident.

The AutoCM is characterized by a three-layer architecture: an Input layer, where the signal is captured from the environment, a Hidden layer, where the signal is

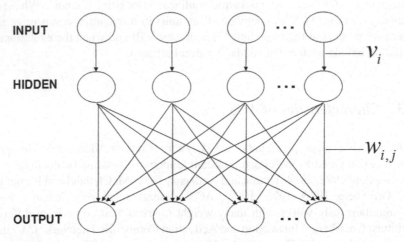

Fig. 1.1 An example of an AutoCM with N = 4

modulated inside the AutoCM, and an Output layer through which the AutoCM feeds back upon the environment on the basis of the stimuli previously received and processed (Fig. 1.1).

Each layer contains an equal number of N units, so that the whole AutoCM is made of *3N* units. The connections between the Input and the Hidden layers are mono-dedicated, whereas the ones between the Hidden and the Output layers are fully saturated, i.e. at maximum gradient. Therefore, given N units, the total number of the connections, Nc, is given by:

$$Nc = N(N + 1)$$

All of the connections of AutoCM may be initialized either by assigning a same, constant value to each, or by assigning values at random. The best practice is to initialize all the connections with a same, positive value, close to zero.

The learning algorithm of AutoCM may be summarized in a sequence of four characteristic steps:

1. Signal transfer from the input into the hidden layer;
2. Adaptation of the values of the connections between the Input and the Hidden layers;
3. Signal transfer from the hidden into the output layer;
4. Adaptation of the value of the connections between the Hidden and the Output layers.

Notice that steps 2 and 3 may take place in parallel.

We write as $m^{[s]}$ the units of the Input layer (sensors), scaled between 0 and 1; as $m^{[h]}$ the units of the Hidden layer, and as $m^{[t]}$ the units of the Output layer (system target). We moreover define **v**, the vector of mono-dedicated connections; **w**, the matrix of the connections between the Hidden and the Output layers; and n, the discrete time that spans the evolution of the AutoCM weights, or, put another way, the number of cycles of processing, counting from zero and stepping up one unit at each completed round of computation: $n \in T$.

In order to specify steps 1–4 that define the AutoCM algorithm, we have to define the corresponding signal forward-transfer equations and the learning equations as follows:

(a) Signal transfer from the Input to the hidden layer:

$$m_{i_{(n)}}^{[h]} = m_i^{[s]}\left(1 - \frac{v_{i_{(n)}}}{C}\right) \tag{1.2}$$

where C is a positive real number not less than 1, which we will refer to as the contraction parameter (see below for comments), and where the (n) subscript has been omitted from the notation of the input layer units, as these remain constant at every cycle of processing. It is useful to set $C = \sqrt[2]{N}$, where N is the number of variables considered.

(b) Adaptation of the connections $v_{i_{(n)}}$ through the variation $\Delta v_{i_{(n)}}$ which amounts to trapping the energy difference generated according to (1.2):

$$\Delta v_{i_{(n)}} = \left(m_i^{[s]} - m_{i_{(n)}}^{[h]}\right) \cdot \left(1 - \frac{v_{i_{(n)}}}{C}\right) \tag{1.3}$$

$$v_{i_{(n+1)}} = v_{i_{(n)}} + \alpha \cdot \Delta v_{i_{(n)}} \tag{1.4}$$

(c) Signal transfer from the hidden to the output layer:

$$Net_{i_{(n)}} = \sum_{j=1}^{N} m_{j_{(n)}}^{[h]} \cdot \left(1 - \frac{w_{i,j_{(n)}}}{C}\right) \tag{1.5}$$

$$m_{i_{(n)}}^{[t]} = m_{i_{(n)}}^{[h]} \cdot \left(1 - \frac{Net_{i_{(n)}}}{C}\right); \tag{1.6}$$

(d) Adaptation of the connections $w_{i,j_{(n)}}$ through the variation $\Delta w_{i,j_{(n)}}$ which amounts, accordingly, to trapping the energy difference as to (1.6):

$$\Delta w_{i,j_{(n)}} = \left(m_{i_{(n)}}^{[h]} - m_{i_{(n)}}^{[t]}\right) \cdot \left(1 - \frac{w_{i,j_{(n)}}}{C}\right) \cdot m_{j_{(n)}}^{[h]} \tag{1.7}$$

$$w_{i,j_{(n+1)}} = w_{i,j_{(n)}} + \alpha \cdot \Delta w_{i,j_{(n)}} \tag{1.8}$$

There are a few important peculiarities of Auto-CMs with respect to more familiar classes of ANNs that need special attention and call for careful reflection:

- Auto-CMs are able to learn when starting from initializations, in which all connections are set to the same value, i.e., they do not suffer the problem of symmetric connections.
- During the training process, Auto-CMs always assign positive values to connections. In other words, Auto-CMs do not allow for inhibitory relations among nodes, but only for different strengths of excitatory connections.
- Auto-CMs can learn in difficult conditions, namely, when the connections of the main diagonal of the second layer connection matrix are removed. In the context of this kind of learning process, Auto- CMs seem to reconstruct the relationship occurring between each pair of variables. Consequently, from an experimental point of view, it seems that the ranking of its connections matrix translates into the ranking of the joint probability of occurrence of each pair of variables.
- Once the learning process has occurred, any input vector belonging to the training set will generate a null output vector. So, the energy minimization of the training vectors is represented by a function through which the trained connections completely absorb the input training vectors. Thus, AutoCM seems to learn how to transform itself in a 'dark body'.
- At the end of the training phase ($\Delta w_{i,j} = 0$), all the components of the weights vector **v** attain the same value:

$$\lim_{n \to \infty} v_{i_{(n)}} = C \tag{1.9}$$

The matrix **w**, then, represents the AutoCM knowledge about the whole dataset.

The AutoCM connections matrix filtered by Minimum Spanning Tree generates an interesting graph whose biological evidence has already been tested in the medical field (Buscema and Grossi 2008; Buscema et al. 2008a, b; Licastro et al. 2010).

Practically, this means that the AutoCM algorithm is able to discover variable similarities completely embedded in the dataset and invisible to the other classic tools. This approach highlights affinities among variables as related to their dynamical interaction rather than to their simple contingent spatial position. This approach describes a context typical of living systems in which a continuous time dependent complex change in the variable value is present. After the training phase, the matrix of the AutoCM represents the warped landscape of the dataset. We apply a simple filter (minimum spanning tree) to the matrix of the AutoCM system to show the map of main connections between and among variables and the principal hubs of the system. These hubs can also be defined as variables with the maximum number of connections in the map.

The AutoCM algorithms used for all the elaborations presented in this chapter are implemented only in Semeion proprietary research software that is available for academic purposes only (Buscema 2000a, b, c; Massini 2007).

We analyzed the brain segment volume changes in the two groups of children with and without IQ impairment condensing each segment of data referring to the left and right sides to see if the Auto-CM could depict a pattern of connections consistent with the different outcomes in cognitive function

1.5 Results

1.5.1 Classification Results

With the data randomly separated into two groups of equal sizes, and thus "blind" as to any predetermined categorization, the following unnumbered tables indicate the results of the four experiments:

Experiment 1	$V_1(\%)$	$V_2(\%)$	A. Mean (%)	W. Mean (%)	Error
Meta-Fuzzy	90.00	92.86	91.43	91.38	2.5
Majority Vote	93.34	85.72	89.52	89.66	3
AQV1	76.67	92.86	84.76	84.49	4.5
B_NRC	86.67	78.58	82.62	82.76	5
B_SCM	83.34	75.00	79.17	79.31	6
SN	66.67	89.29	77.98	77.59	6.5
B_AutoBP	80.00	75.00	77.50	77.59	6.5
B_AutoCM	73.33	71.43	72.38	72.41	8
BP	73.33	67.86	70.60	70.69	8.5
Mean	80.37	80.95	80.66	80.65	5.61

Experiment 2	$V_1(\%)$	$V_2(\%)$	A. Mean (%)	W. Mean (%)	Error
Meta-Fuzzy	83.34	82.14	82.74	82.76	5.00
B_SCM	76.67	85.71	81.19	81.04	5.50
AQV1	76.67	85.71	81.19	81.04	5.50
Majority Vote	83.34	78.57	80.95	81.04	5.50
SN	66.67	85.71	76.19	75.86	7.00
B_NRC	90.00	53.57	71.79	72.42	8.00
B_AutoCM	63.34	82.14	72.74	72.41	8.00
B_AutoBP	70.00	67.86	68.93	68.97	9.00
BP	76.67	50.00	63.34	63.80	10.50
Mean	76.30	74.60	75.45	75.48	7.11

Experiment 3	$V_1(\%)$	$V_2(\%)$	A. Mean (%)	W. Mean (%)	Error
Majority Vote	96.67	89.29	92.98	93.11	2.00
Meta-Fuzzy	93.34	89.29	91.31	91.38	2.50
SN	83.34	89.29	86.31	86.21	4.00
B_SCM	90.00	71.43	80.72	81.04	5.50
B_NRC	83.34	75.00	79.17	79.31	6.00
AQV1	83.34	75.00	79.17	79.31	6.00
B_AutoBP	76.67	78.57	77.62	77.59	6.50
B_AutoCM	70.00	75.00	72.50	72.41	8.00
BP	73.34	67.86	70.60	70.69	8.50
Mean	83.33	78.97	81.15	81.23	5.44

Experiment 4	V$_1$(%)	V$_2$ (%)	A. Mean (%)	W. Mean (%)	Error
SN	93.33	85.71	89.52	89.66	3.00
Majority Vote	90.00	82.14	86.07	86.21	4.00
B_NRC	83.33	85.72	84.53	84.49	4.50
Meta-Fuzzy	86.67	82.14	84.41	84.49	4.50
BP	73.34	89.29	81.31	81.04	5.50
B_SCM	80.00	78.57	79.29	79.31	6.00
B_AutoBP	83.34	71.43	77.38	77.59	6.50
B_AutoCM	93.34	60.72	77.03	77.59	6.50
AQV1	73.34	82.14	77.74	77.59	6.50
Mean	84.07	79.76	81.92	81.99	5.22

1.5.2   Interpretation of the Experiments

The results from all of these experiments are incredibly good and thus demonstrates the existence of a robust non-linear link between the radiation treatment and the IQ in each child's brain. The application of very different types of non-linear algorithms, taking care to address the specific capabilities required by each method, permits us to recognize and correctly categorize the children with varying degrees of brain damage.

The results from experiment 2 are the worst of the group and hence, the original dataset is absent of noise. This means that if we permit the data to be transformed such that the volume differences are represented only by the sign difference, critical information is lost. From this experiment we now can assign greater confidence in our results.

Experiments 3 and 4 make it evident that the right and left hemispheres of the brain work in tandem. Each side compensates for the other and their summation preserves key, critical information, and the sign of their summation is not fundamental to understanding the damages resulting from the radiotherapy treatment.

1.6   Prototypes Identification Through ACS

The results of using ACS to identify the prototypes contained in the brain data are shown in Table 1.6.

Figure 1.2 shows the dynamics of activity created by ACS on the dataset for subjects with IQ < 94 and Fig. 1.3 shows the dynamics for subjects with IQ ≥ 94.

1.7   Discovering Hidden Links with Auto-CM

The graph of children without IQ impairment shows a connection scheme among brain segments which is consistent with the natural anatomic relation in the brain (Fig. 1.3). The hippocampus acts as the central node and divides the graphs

Table 1.6 Comparison of prototypes identified using ACS

Variables	Activation
Prototype of subject with IQ < 94	
ACS after 553 cycles	
Cerebral white matter	−0.9990
Cerebral cortex	1.0000
Thalamus	−1.0000
Caudate	−0.9950
Putamen	−0.9980
Pallidum	−1.0000
Globus pallidus	−1.0000
Hippocampus	1.0000
Amygdala	−0.9980
IQ < 94 (external input)	1.0000
IQ ≥ 94	−1.0000
Prototype of subject with IQ ≥ 94	
ACS after 1,352 cycles	
Cerebral white matter	1.0000
Cerebral cortex	−0.9270
Thalamus	1.0000
Caudate	−0.0750
Putamen	0.9460
Pallidum	1.0000
Globus pallidus	1.0000
Hippocampus	−0.8950
Amygdala	1.0000
IQ < 94	−0.8500
IQ ≥ 94 (external input)	1.0000

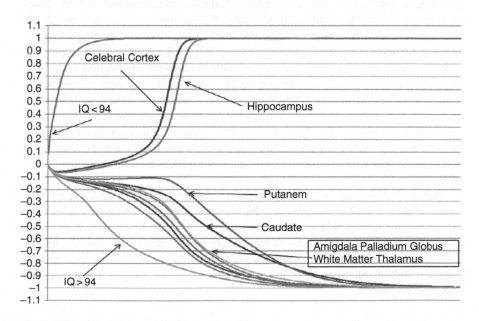

Fig. 1.2 Dynamics of IQ < 94 subject prototype

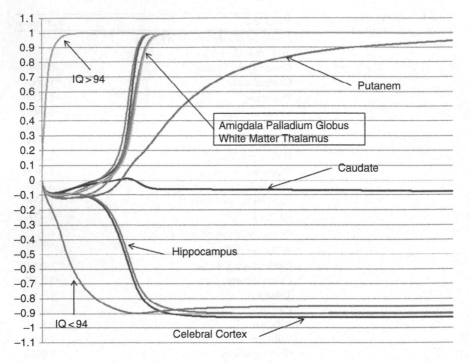

Fig. 1.3 Dynamics of IQ ≥ 94 subject prototype

in two sections: In the upper part of the graph the cerebral cortex is connected with the amygdala and globus pallidus, the latter being linked to putanem. All the values of connections strength among brain segments volume changes are very high. The high connection strength values indicate that the volume changes within the segment of brain more related to cognitive function, like hippocampus (memory), and amygdala (emotions and fluency) are non-linearly closely related each-other suggesting a sort of compensation among them in the eventual volume losses.

In the lower part of the graph in fig. 1.2, the thalamus, caudate, pallidum and cerebral white matter, segments less related to cognitive function, have very low connection strength values. This suggests that in these segments no compensation of eventual volume losses took place but this did not impair IQ. This behavior is consistent with the dynamic trends obtained with ACS analysis in the lower part of Fig. 1.4.

The connection map of the brain segment volume change in children with IQ impairment is completely different. Here, connections do not closely reflect the normal brain anatomy. The graph is less complex, with the thalamus proper acting as a hub. All the values of connections strength among brain segments less related to cognitive function (globus pallidus, pallidum, caudate, white cerebral matter, putamen andamygdala) are very high, while the opposite is true for hippocampus and cerebral cortex.

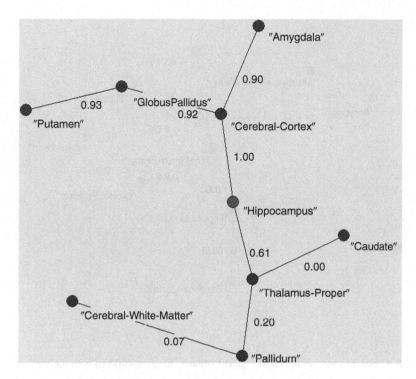

Fig. 1.4 Connection map of the brain segment volume change in children without IQ impairment

The low connection strength values among hippocampus and cerebral cortex indicate that the volume changes within the segment of the brain more related to cognitive function are poorly related to each other suggesting that no compensation took place and therefore, this could explain IQ impairment (Fig. 1.5).

1.8   Comments and Conclusions

The supervised artificial neural networks have allowed us to show that a highly nonlinear relation does exist between brain volumes changes and IQ

The ACS system defines the specific features of two important prototypes: the prototype of the subjects whose IQ, after the radiotherapy treatment, is measured to be less than 94, and the prototype of the subjects whose IQ, after treatment, is measured to be greater than or equal to 94. The IQ < 94 subset seems to be specific to the subjects with a volume alteration focused in left and right parts of the Cerebral Cortex and Hippocampus after the treatment. For these subjects it appears that compensation between left and right hemispheres of the brain seem to be more difficult.

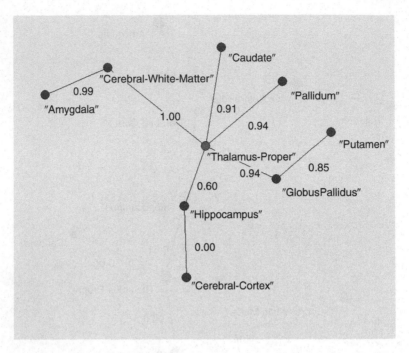

Fig. 1.5 Connection map of the brain segment volume change in children with IQ impairment

The IQ $\geq$ 94 subset seems to be characterized by a more distributed alteration of brain volumes after treatment. White matter, Thalamus, Palladium, Globus Pallidus and Amigdala present small alterations, and the left and the right sides of the Cerebral Cortex and Hippocampus seem to be more preserved. Results consistent with this trend have been obtained with Auto-CM analysis which maps the connection pattern among different brain segments through a nonlinear computation of volume changes before and after radiotherapy.

This initial study lends considerable hope to further study with an expectation that follow-up studies, using the Semeion software (Buscema 2000a, 2000b, 2007a, 2008a, b, 2009c; Massini 2007), will allow physicians to use structural imaging to predict changes in IQ and potentially minimize the adverse effects of radiotherapy treatment on young brains. Additional study based on a dataset of more mature individuals could lend similar support to adversity minimization in adults and perhaps the elderly.

References

Buscema M (1998a) MetaNet: the theory of independent judges. In: Substance use & misuse, vol 33(2), pp 43461 (Models). Marcel Dekker, Inc., New York

Buscema M (1998b) Recirculation neural networks. In: Substance use & misuse, vol 33(2), pp 383–388 (Models). Marcel Dekker, Inc., New York

Buscema M (2000a) Constraints satisfaction networks. Semeion software #14, v. 12.5, Rome, 2000–2009

Buscema M (2000b) Supervised ANNs and organism. Semeion software #12, v. 16, Rome, 2000–2010

Buscema M (2007a) Meta Net. Semeion software no. 44, v.8.0, Rome, 2007–2010

Buscema M (2007b) Squashing theory and contractive map network. Semeion technical paper #32, Rome

Buscema M (2008a) MST. Software for programming graphs from artificial networks. Semeion software #38, Rome, v 6.0, 2008–2009

Buscema M (2008b) Supervised auto-associative ANNs. Semeion software #50, version 3.5, 2008–2010

Buscema M (2009a) Activation and competition system. Mimeo, Semeion, Rome

Buscema M (2009b) Adaptive learning quantization. Mimeo, Semeion, Rome

Buscema M (2009c) Modular auto-associative ANNs, Ver 10.0. Semeion software #51, Rome, 2009–2010

Buscema M, Grossi E (2008) The semantic connectivity map: an adapting self-organizing knowledge discovery method in data bases. Experience in gastro-oesophageal reflux disease. Int J Data Min Bioinf 2(4):362–404

Buscema M, Grossi E, Intraligi M, Garbagna N, Andriulli A, Breda M (2005) An optimized experimental protocol based on neuro-evolutionary algorithms: application to the classification of dyspeptic patients and to the prediction of the effectiveness of their treatment. Artif Intell Med 34:279–305

Buscema M, Breda M, Terzi S (2006a) A feed forward sine based neural network for functional approximation of a waste incinerator emissions. Proceedings of the 8th WSEAS international conference on automatic control, modeling and simulation, Praga

Buscema M, Breda M, Terzi S (2006b) Using sinusoidal modulated weights improve feed-forward neural network performances in classification and functional approximation problems. WSEAS Trans Inf Sci Appl 5(3):885–893

Buscema M, Terzi S, Maurelli G, Capriotti M and Carlei V (2006) The smart library architecture of an orientation portal. In: Quality & quantity. Springer, Netherland, vol 40, pp 911–933

Buscema M, Grossi E, Snowdon D, Antuono P (2008a) Auto-contractive maps: an artificial adaptive system for data mining: an application to Alzheimer disease. Curr Alzheimer Res 5:481–498

Buscema M, Helgason C and Grossi E (2008) Auto contractive maps, H function and maximally regular graph: theory and applications. Special session on Artificial adaptive systems in medicine: applications in the real world, NAFIPS 2008 (IEEE), New York

Chauvin Y, Rumelhart DE (eds) (1995) Backpropagation: theory, architectures, and applications. Lawrence Erlbaum Associates, Inc. Publishers, New Jersey

Dale AM, Sereno MI (1993) Improved localization of cortical activity by combining EEG and MEG with MRI cortical surface reconstruction: a linear approach. J Cogn Neurosci 5(2):162–176

Dale AM, Fischl B, Sereno MI (1999) Cortical surface-based analysis I: segmentation and surface reconstruction. Neuroimage 9:179–194

Day WHE (1988) Consensus methods as tools for data analysis. In: Bock HH (ed) Classification and related methods for data analysis. North-Holland, Amsterdam, pp 312–324

Diappi L, Bolchim P, Buscema M (2004) Improved understanding of urban sprawl using neural networks. In: Van Leeuwen JP, Timmermans HJP (eds) Recent advances in design and decision support systems in architecture and urban planning. Kluwer Academic Publishers, Dordrecht

Fischl B, Salat D, Busa E, Albert M, Dieterich M, Haselgrove C, van der Kouwe A, Killiany R, Kennedy D, Klaveness S, Montillo A, Makris N, Rosen B, Dale AM (2002) Whole brain segmentation. Automated labeling of neuroanatomical structures in the human brain. Neuron 33(3):341–355

Fischl B, Salat DH, van der Kouwe AJ, Makris N, Segonne F, Quinn BT, Dale AM (2004) Sequence-independent segmentation of magnetic resonance images. Neuroimage 23(Suppl 1): S69–S84

Grossberg S (1976) Adaptive pattern classification and universal recording: Part I. Parallel development and coding of neural feature detectors. Biol Cybern 23:121–134

Grossberg S (1978) A theory of visual coding, memory, and development. In: Leeuwenberg J, Buffart HFJ (eds) Formal theories of visual perception. Wiley, New York

Grossberg S (1981) How does the brain build a cognitive code? Psychol Rev 87:1–51

Harila-Saari AH, Paakko EL, Vainionpaa LK, Pyhtinen J, Lanning BM (1998) A longitudinal magnetic resonance imaging study of the brain in survivors in childhood acute lymphoblastic leukemia. Cancer 83(12):2608–2617

Hertzberg H, Huk WJ, Ueberall MA, Langer T, Meier W, Dopfer R, Skalej M, Lackner H, Bode U, Janssen G, Zintl F, Beck JD (1997) CNS late effects after ALL therapy in childhood. Part I: Neuroradiological findings in long-term survivors of childhood ALL – an evaluation of the interferences between morphology and neuropsychological performance. The German Late Effects Working Group. Med Pediatr Oncol 28(6):387–400

Hinton GE, Anderson A (eds) (1981) Parallel models of associative memory. Erlbaum, Hillsdale, NJ

Hopfield JJ (1982) Neural networks and physical systems with emergent collective computational abilities. Proc Natl Acad Sci USA 79:2554–2558

Hopfield JJ (1984) Neurons with graded response have collective computational properties like those of two-state neurons. Proc Natl Acad Sci USA 81:3088–3092

Ishikawa N, Tajima G, Yofune N, Nishimura S, Kobayashi M (2006) Moyamoya syndrome after cranial irradiation for bone marrow transplantation in a patient with acute leukemia. Neuropediatrics 37(6):364–366

Khong PL, Leung LH, Fung AS, Fong DYT, Qiu D, Kwong DLW, Ooi GC, McAlanon G, Cao G, Chan GCF (2006) White matter anisotropy in post-treatment childhood cancer survivors: preliminary evidence of association with neurocognitive function. J Clin Oncol 24(6):884–890

Kikuchi A, Maeda M, Hanada R, Okimoto Y, Ishimoto K, Kaneko T, Ikuta K, Tsuchida M (2007) Moyamoya syndrome following childhood acute lymphoblastic leukemia. Pediatr Blood Cancer 48(3):268–272

Kohonen T (1995) Self-organizing maps. Springer Verlag, Berlin

Kosko B (1992) Neural networks for signal processing. Prentice Hall, Englewood Cliffs, NJ

Kuncheva LI (2004) Combining pattern classifiers: methods and algorithms. Wiley, Hoboken, NJ

Laitt RD, Chambers EJ, Goddard PR, Wakeley CJ, Duncan AW, Foreman NK (1995) Magnetic resonance imaging and magnetic resonance angiography in long term survivors of acute lymphoblastic leukemia treated with cranial irradiation. Cancer 76(10):1846–1852

Leung LH, Ooi GC, Kwong DL, Chan GC, Cao G, Khong PL (2004) White-matter diffusion anisotropy after chemo-irradiation: a statistical parametric mapping study and histogram analysis. Neuroimage 21(1):261–268

Licastro F, Porcellini E, Chiappelli M, Forti P, Buscema M, Ravaglia G, Grossi E (2010) Multivariable network associated with cognitive decline and dementia. Int Neurobiol Aging 1(2):257–269

Liu AK, Marcus KJ, Fischl B, Grant PE, Poussaint TY, Rivkin MY, Davis P, Tarbell NJ, Yock TI (2007) Changes in cerebral cortex of children treated for medulloblastoma. Int J Radiat Oncol Biol Phys 68(4):992–998

Massini G (2007) Semantic connection map, Ver 2.0. Semeion software #45, Rome, 2007–2009

McClelland JL, Rumelhart DE (1988) Explorations in parallel distributed processing: a handbook of models, programs, and exercises. Bradford Books, Cambridge, MA

Mulhern RK, Reddick WE, Palmer SL, Glass JO, Elkin TD, Kun LE, Taylor J, Langston J, Gajjar A (1999) Neurocognitive deficits in medulloblastoma survivors and white matter loss. Ann Neurol 46(6):834–841

Nagel BJ, Palmer SL, Reddick WE, Glass JO, Helton KJ, Wu S, Xiong X, Kun LE, Gajjar A, Mulhern RK (2004) Abnormal hippocampal development in children with medulloblastoma treated with risk-adapted irradiation. AJNR Am J Neuroradiol 25(9):1575–1582

Paakko E, Talvensaari K, Pyhtinen J, Lanning M (1994) Late cranial MRI after cranial irradiation in survivors of childhood cancer. Neuroradiology 36(8):652–655

Paakko E, Harila-Saari A, Vanionpaa L, Himanen S, Pyhtinen J, Lanning M (2000) White matter changes on MRI during treatment in children with acute lymphoblastic leukemia: correlation with neuropsychological findings. Med Pediatr Oncol 35(5):456–461

Poussaint TY, Siffert J, Barnes PD, Pomeroy SL, Goumnerova LC, Anthony DC, Sallan SE, Tarbell NJ (1995) Hemorrhagic vasculopathy after treatment of central nervous system neoplasia in childhood: diagnosis and follow-up. AJNR Am J Neuroradiol 16(4):693–699

Reddick WE, White HA, Glass JO, Wheeler GC, Thompson SJ, Gajjar A, Leigh L, Mulhern RK (2003) Developmental model relating white matter volume to neurocognitive deficits in pediatric brain tumor survivors. Cancer 97(10):2512–2519

Reddick WE, Glass JO, Palmer SL, Wu S, Gajjar A, Langston LW, Kun LE, Xiong X, Mulhern RK (2005) Atypical white matter volume development in children following craniospinal irradiation. Neuro Oncol 7(1):12–19

Reddick WE, Shan ZY, Glass JO, Helton S, Xiong X, Wu S, Bonner MJ, Howard SC, Christensen R, Khan RB, Pui CH, Mulhern RK (2006) Smaller white-matter volumes are associated with larger deficits in attention and learning among long-term survivors of acute lymphoblastic leukemia. Cancer 106(4):941–949

Rumelhart DE, McClelland JL (eds) (1986) Parallel distributed processing, Vol. 1. Foundations, explorations in the microstructure of cognition, Vol. 2. Psychological and biological models. The MIT Press, Cambridge, MA

Rumelhart DE, Smolensky P, McClelland JL, Hinton GE (1986) Schemata and sequential thought processes in PDP models. In: McClelland JL, Rumelhart DE (eds) PDP, exploration in the microstructure of cognition, vol II. The MIT Press, Cambridge, MA

Segonne F, Dale AM, Busa E, Glessner M, Salat D, Kahn HK, Fischl B (2004) A hybrid approach to the skull stripping problem in MRI. Neuroimage 22(3):1060–1075

Ullrich NJ, Robertson R, Kinnamon DD, Scott RM, Kieran MW, Turner CD, Chi SN, Goumnerova L, Proctor M, Tarbell NJ, Marcus KJ, Pomeroy SL (2007) Moyamoya following cranial irradiation for primary brain tumors in children. Neurology 68(12):932–938

Chapter 2
J-Net: An Adaptive System for Computer-Aided Diagnosis in Lung Nodule Characterization

Massimo Buscema, Roberto Passariello, Enzo Grossi, Giulia Massini, Francesco Fraioli, and Goffredo Serra

2.1 Introduction

Lung cancer is the leading cause of cancer deaths in the western world, with a total number of deaths greater than that resulting from colon, breast, and prostate cancers combined (Greenlee et al. 2000). The appearance of a non-calcified solitary lung nodule on a chest radiograph or CT, often serendipitous, is the most common diagnostic sign of lung cancer. Currently, a significant research effort is being devoted to the detection and characterization of lung nodules on thin-section computed tomography (CT) images. This represents one of the newest directions of CAD development in thoracic imaging.

At the present time the Multi Detector Computed Tomography (MDCT) is the gold standard in the detection of lung nodules (Henschke and Yankelevitz 2008; Diederich et al. 2002, 2003; Henschke et al. 2002; Swensen et al. 2003; Fischbach et al. 2003); it is well demonstrated that early lung cancer often occurs as a small

M. Buscema (✉)
Semeion Research Center of Sciences of Communication, Via Sersale 117, Rome, Italy

Department of Mathematical and Statistical Sciences, CCMB, University of Colorado, Denver, Colorado, USA
e-mail: m.buscema@semeion.it

R. Passariello
Department of Radiological Sciences, University of Rome, "La Sapienza", Rome, Italy

E. Grossi
Bracco SpA Medical Department, San Donato Milanese, Italy

G. Massini
Semeion Research Centre, Via Sersale n. 117, 00128 Rome, Italy

F. Fraioli • G. Serra
Department of Radiological Sciences, University of Rome, "La Sapienza", Rome, Italy

W.J. Tastle (ed.), *Data Mining Applications Using Artificial Adaptive Systems*,
DOI 10.1007/978-1-4614-4223-3_2, © Springer Science+Business Media New York 2013

undefined precancerous lung nodule (PN) (Li et al. 2002), although only a few in number actually result in lung cancers.

In most institutions MDCT follow-up remains the most common approach in the differential diagnosis for nodules smaller than 1 cm; unfortunately this procedure is a substantial source of patient anxiety, radiation exposure, and medical cost because of the number of resultant follow-up scans.

In a screening program with CT, the radiologist has to deal with a large number of images and therefore detection errors (failure to detect a cancer) or interpretation errors (failure to correctly diagnose a detected cancer) can occur (Li et al. 2002). In such a circumstance, a CAD scheme for detection and for characterization of lung nodules would be particularly useful for the reduction of detection errors and interpretation errors, respectively. In particular a computerized characterization scheme can provide quantitative information such as the likelihood of malignancy to assist radiologists in diagnosing a detected nodule (Diederich et al. 2002).

Some authors investigated the use of Computer-Aided Diagnosis (CAD) systems to classify malignant and benign lung nodules found on CT scans (Aoyama et al. 2003a, b; Shiraishi et al. 2006; Goldin et al. 2008). These investigations showed, generally speaking, promising results and supported the idea that CAD programs can improve the radiologist's diagnostic efficiency. The methodology underlying CAD for lung nodules characterization is generally based on the extraction of a definite set of features from the segmented nodule image and also from the outside region based on 2D sectional data and 3D volumetric data. These features represent the inputs to linear or nonlinear classifiers for distinguishing between benign and malignant nodules (Shiraishi et al. 2003; Li 2007).

In this chapter we describe a new CAD system based on a completely new Artificial Neural Network (ANN) algorithms created for image enhancement and analysis:

1. Active Connection Fusion (ACF): a new set of ANNs for image fusion (Software (Buscema 2010));
2. J-Net Active Connections Matrix (J-Net): a new ANN for dynamic image segmentation. Patent (Buscema 2003, 2004, 2007); Software (Buscema 2003–2010);
3. Population (Pop): a new and fast multidimensional scaling algorithm able to squash hyper-points from a high dimensional space onto a small dimensional space with minimal deformations (Software (Massini 2007–2009);
4. Adaptive Learning Quantization (AVQ) and Meta-Consensus: two new supervised ANNs, experts in rapid classification and not sensitive to over fitting (Software (Buscema 1999–2010, 2008–2010).

These algorithms pre-process the images of the PNs obtained from the MDCT and, consequently, make the following processes of segmentation easier, and shape feature extraction and diagnosis. The aim is to detect and measure the small densitometric differences at the closest periphery and in the inner regions of a nodule not visible to the human eye.

The purpose of this chapter is to verify whether the extracted shape features could be used to differentiate benign lesions from malignant ones.

2.2   Materials and Methods

2.2.1   Patients

We included in our analysis 88 patients with 90 nodules (two patients had two nodules; mean age 65 years, 49–83 years) with a solitary undefined lung nodule less than or equal to 20 mm; Forty four of them had benign nodules (34 males and 10 females) and 46 malignant nodules (30 males and 17 females), all primary neoplasms.[1] The average diameter of benign and malignant nodules was equal to 15.8 mm and 17.2 mm respectively.

All the patients were included in our study after an annual high resolution CT follow up. Those patients with an increase in size of the PN were considered as being suspicious of malignancy. In these patients a transthoracic needle biopsy ($n = 15$) or surgery ($n = 30$), were performed to confirm the diagnosis. For one patient unavailable for surgical procedure, a positive PET CT was accepted for diagnosis (SUV 4). The remainder of the patients with a stable size PN at the annual follow up were considered as benign. Eligibility criteria are determined by the health professionals collecting the data.

2.2.2   CT-Investigation

The CT system was a Siemens Somatom Sensation Cardiac (Siemens, Enlargen, Germany). The CT examination was performed on a 64 MDCT using a thin collimation (0.6 mm) protocol. Exposure parameters were 100 mA, 120 kV; gantry rotation time was 0.33 s and scan time was 4 s. No contrast media was administered. The CT slice thickness was 1.5 mm; all images were reconstructed by using a bone algorithm (B60).

2.2.3   Data

The data recorded for each lesion are:

1. A set of consecutive images representing the lesion: this sequence of images was selected from the CT Image analysis by the experts. The images are in BMP format; dimension variation is between 556×800 and $1,196 \times 1,357$ pixels;
2. The malignancy or benignancy of the lesion;

---

[1] The 90 CT volumes were provided by the Department of Radiological Sciences of the University of Rome, "La Sapienza".

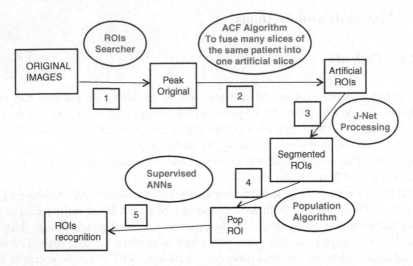

Fig. 2.1 Images processing system fusion and J-Net based

3. The position of the lesion on the sequence of images extracted from the CT analysis. For this parameter we considered the (x, y) coordinates of the center point of the lesion (as evaluated by the experts).

Additional information like the expert radiologist's diagnosis as well as the size (maximum diameter) of each lesion were also stored, but not used, in the analysis.

2.3 The Processing System

A complex system composed of different steps and sub processing systems is necessary in order to analyze the assigned dataset images (Fig. 2.1).

1. The Dataset of images is decomposed into a dataset of regions of interest (ROI): each ROI is a Rows x Columns box centered on a specific tumor, benign or malignant. Each lesion is represented by a different number of ROIs, because it is defined by a specific number of slices. We have renamed these new images Original ROIs.[2]
2. In the second step we use a new ANN, the *Active Connection Fusion* (ACF), to fuse the different images (ROIs) of the same lesion into only one new image (ROI), containing all the key information of the 3D lesion. ACF is a new ANN able to fuse many different and registered images onto one image conserving and

---

[2] Software for extracting ROIs from the original images was set up by Dr. Petritoli and Dr. Terzi (Semeion, Research Center).

even enhancing the most important features of the source images. We have renamed these new images Artificial ROIs.

3. The third step of this process is executed by the J-Net Algorithm. J-Net is a special system able to determine the *shapes* and the *skeleton* of an image at different levels of light intensity. Consequently, J-Net generates for each single artificial ROI a set of new images, each one with a shape and skeleton resulting from different light intensities from images with shape and skeleton of the original ROI detected at very low light intensity, up to the images whose shapes and skeleton is detected where the light intensity of the original ROI is very high. The main goal of the J-Net system is to find the main features of any assigned image. We have renamed these new images J-Net ROIs.

4. The fourth step is the generation of the Histogram of each J-Net ROI: each J-Net ROI is coded into a 256 input vector in which each vector component is coded with a number of presences derived from any grey tone, onto each lesion. Because we have set the processing of each lesion with five different alpha values, J-Net generates five images for each lesion. Consequently, we have re-coded each lesion into 1,280 inputs (256×5). These new outputs are named Histogram ROIs.

5. A new Multidimensional Scaling Algorithm, named Population, will squash each huge vector (the Histogram of 1,280 components for each lesion) into a more compact vector representing the main features of each original ROI. The Population algorithm is discussed below. These new compact vectors, generated by Population, are named Pop ROIs.

This new dataset composed of all the squashed vectors (Pop ROIs) will be analyzed using different supervised learning algorithms. A Five K-Fold Cross Validation protocol is used to analyze the results of the pattern recognition process. Two new supervised ANNs will be presented and compared against other more standard Learning Machines and ANNs.

2.4   The Active Fusion Matrix Algorithm

Scientific literature about Image Fusion is quite considerable (Blum and Liu 2006), especially in the multisensory military field. Multi-sensor fusion refers to the direct combination of several signals in order to provide a signal that has the same general format as the source signals. Consequently, image fusion generates a fused image in which each pixel is determined from a set of pixels in each source image. We present a new image fusion algorithm named Active Connection Fusion (ACF) and we compare it with the best algorithms used in literature. Then we use ACF to fuse the different slices of the same lesion into one artificial ROI. This new artificial ROI should preserve the key information about the lesion distributed in the different original slices. The advantage is clear:

1. We work using only one image per lesion, without considering the various slices that are representative of each lesion, and
2. We preserve the most important features of each lesion.

2.4.1 Active Connection Fusion: Logic and Equations

The ACF system is composed of a series of analytical steps:

- Linear stretching;
- Non linear focusing;
- Weights initialization;
- Delta calculation;
- Weights update;
- Fused image visualization.

During this process ACF transforms many source images into one new image. The purpose of ACF is to preserve the most important features and details of the source images and place them in the new artificial image.

Linear Stretching:

Legend :
$$P(0)_{i,j}^m = \text{ source pixel(x,y) of the m-th source.}$$

$$P(1)_{i,j}^m = F\left(P(0)_{i,j}^m\right) = Scale^m \cdot P(0)_{i,j}^m \cdot Offset^m; \ P(1)_{i,j}^m \in [0,1]. \tag{2.1}$$

Non Linear Focusing:

Legend :
$$M = \text{ Number of Source Images;}$$

$$\bar{x}_{i,j} = \frac{1}{M} \cdot \sum_m^M P(1)_{i,j}^m; m \in [1, 2, ..., M]. \tag{2.2}$$

$$c_{i,j} = \frac{\bar{x}_{i,j}}{1.0 - \bar{x}_{i,j}}; \tag{2.3}$$

$$d_{i,j}^m = P(1)_{i,j}^m - \bar{x}_{i,j}; \tag{2.4}$$

$$P(2)_{i,j}^m = \frac{c_{i,j}}{c_{i,j} + e^{-d_{i,j}^m}}. \tag{2.5}$$

Weights Initialization:

Legend :

$$R = \text{ Radius of Pixel Neighborhood.}$$

$$\sigma_{i,j}^m = \sum_{k=-R}^{R} \sum_{z=-R}^{R} \left(P(2)_{i,j}^m - P(2)_{i+k,j+z}^m \right)^2 ; \qquad (2.6)$$

$$Win = \underset{m}{ArgMax} \left\{ \sigma_{i,j}^m \right\}; \quad m \in [1, 2, ..., M]; \qquad (2.7)$$

$$W_{i,j,i+k,j+z}^{(t)} = 2.0 \cdot P(2)_{i,j}^{Win} - 1.0; \quad k, z \in [-R, +R]. \qquad (2.8)$$

Delta Calculation:
Each pixel of any image versus its neighborhood:

$$\alpha_{i,j,i+k,j+z} = \frac{1}{M} \cdot \sum_{m}^{M} \left(P(2)_{i,j}^m - P(2)_{i+k,j+z}^m \right)^2 ; \qquad (2.9)$$

Each neighborhood of an image versus each neighbor of the other images:

$$\beta_{i,j,i+k,j+z} = \frac{2}{M \cdot (M-1)} \cdot \sum_{n \neq}^{M-1} \sum_{m}^{M} \left(P(2)_{i+k,j+z}^n - P(2)_{i+k,j+z}^m \right)^2 ; \qquad (2.10)$$

Each central pixel of each image versus the neighbor of the other images:

$$\gamma_{i,j,i+k,j+z} = \sqrt[2]{\frac{2}{M \cdot (M-1)} \cdot \sum_{n \neq}^{M-1} \sum_{m}^{M} \left(P(2)_{i,j}^n - P(2)_{i+k,j+z}^m \right)^2} ; \qquad (2.11)$$

$$\delta_{i,j,i+k,j+z} = \sqrt[2]{\frac{2}{M \cdot (M-1)} \cdot \sum_{n \neq}^{M-1} \sum_{m}^{M} \left(P(2)_{i+k,j+z}^n - P(2)_{i,j}^m \right)^2} ; \qquad (2.12)$$

$$\varphi_{i,j,i+k,j+z} = \sqrt[2]{\alpha_{i,j,i+k,j+z} + \beta_{i,j,i+k,j+z}} ; \qquad (2.13)$$

$$\phi_{i,j,i+k,j+z} = \gamma_{i,j,i+k,j+z} \cdot \delta_{i,j,i+k,j+z}. \qquad (2.14)$$

Weights update:

$$\psi_{i,j,i+k,j+z} = \frac{\varphi_{i,j,i+k,j+z}}{\phi_{i,j,i+k,j+z}} ; \qquad (2.15)$$

$$y = \psi_{i,j,i+k,j+z} ; \qquad (2.15a)$$

$$W_{i,j,i+k,j+z}^{(t+1)} = W_{i,j,i+k,j+z}^{(t)} + \frac{e^{-y} - e^{y}}{e^{-y} + e^{y}}. \tag{2.16}$$

Visualization:

$$NewP_{i,j} = f\left(\frac{1}{(2 \cdot R + 1)^2} \cdot \sum_{k=-R}^{R} \sum_{z=-R}^{R} W_{i,j,i+k,j+z}^{(t)}\right); \tag{2.17}$$

$$f(\cdot) = NewP_{i,j} \in [0, 255]; \quad \text{Linear Stretching}. \tag{2.18}$$

2.5   Active Connection Fusion: Application and Comparisons

We have tested ACF with many images and we have compared the ACF algorithm with different fusion algorithms known in the literature. Here we present a small set of examples in which we match ACF with one of the best fusion algorithms actually used, Wavelet.

In Figs. 2.2 and 2.3 there are two x-ray images of a desktop, one taken with high energy and the other taken with low energy. The target in this field is to preserve the

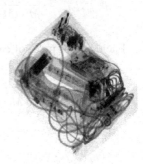

Fig. 2.2 Desktop high energy

Fig. 2.3 Desktop low energy

Fig. 2.4 Desktop wavelet

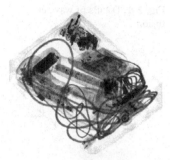

Fig. 2.5 Desktop ACF

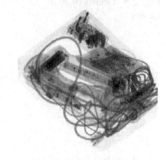

Fig. 2.6 Details of wavelet
fusion

Fig. 2.7 Details of ACF
fusion

penetration of high x-ray energy and the sensitivity towards the detail contained in
the low x-ray energy.

In Figs. 2.4 and 2.5 the Wavelet and the ACF algorithms are compared, and in
Figs. 2.5, 2.6, 2.7, 2.8 and 2.9 the details of the two different fusion algorithms are shown.

Fig. 2.8 Details of wavelet
fusion

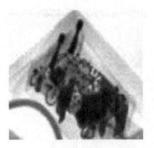

Fig. 2.9 Details of ACF
fusion

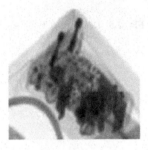

Fig. 2.10 Baggage – high
energy

It is evident how ACF fusion is much more informative than the Wavelet algorithm
in terms of detail preservation, noise elimination and global image enhancement.
Figures 2.10 and 2.11 show another example of x-rays at low and high energy.
Figures 2.12 and 2.13 show the different fusion processing of ACF and Wavelet.

In this example the ACF algorithm shows itself to be considerably more effective than Wavelet.

In the field of security the rapid fusion of images coming from infrared and the
visible band is very useful. Figures 2.14 and 2.15 show the same scene in two
different modalities, infrared and TV. Figure 2.16 is the fusion generated by ACF.

Fig. 2.11 Baggage – low energy

Fig. 2.12 Baggage – wavelet fusion

Fig. 2.13 Baggage – ACF

Fig. 2.14 Infrared

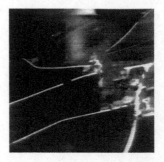

Fig. 2.15 TV

Fig. 2.16 ACF fusion

Figures 2.14, 2.15 and 2.16 show the same problem of infrared and TV image fusion. We compare the ACF solution with the standard solution actually adopted (Figs. 2.17 and 2.18).

The comparison between two fused images (2.19 and 2.20) shows that the image fusion generated by the ACF algorithm evidences a greater level of details.

In this chapter we are interested only in showing the effectiveness of the ACF algorithm in the medical field and in particular, to its application to MDCT. Therefore Figs. 2.21 and 2.22 show the fusion of different slices of a benign and of a malignant tumor into two artificial images.

Fig. 2.17 Infrared

Fig. 2.18 TV image

Fig. 2.19 Standard fusion

Fig. 2.20 ACF fusion

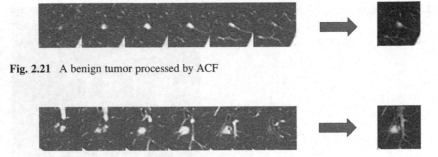

Fig. 2.21 A benign tumor processed by ACF

Fig. 2.22 A malignant tumor processed by ACF

2.6 ACM Systems

The Active Connections Matrix (ACM) [(Buscema et al. 2006), Patent (Buscema 2003, 2004] systems are a new collection of unsupervised artificial adaptive systems developed by Buscema at Semeion Research Institute (Buscema et al. 2006). They were created for automatically extracting features of interest (e.g. edges, segmentation, tissue differentiation, etc.) from digital images. Their main task is the selection of the local properties of interest through a reduction of image noise while maintaining the spatial resolution of high contrast structures and the expression of hidden morphological features. We could formally define an ACM system as a nonlinear adaptive filter based on *local*, *deterministic* and *iterative* operations:

- *Local*, because in each elaboration cycle the operations involve a central pixel and its relations with the very contiguous pixels (the neighborhood of the central pixel).

- *Deterministic*, because the static state towards which the dynamic system tends is represented by the matrix of pixels with the new image based on deterministic equations. Therefore the elaboration can always be repeated resulting in the same outcome.
- *Iterative*, because the operations of the dynamic system repeat themselves, iteratively, until the evolution in the space of phases reaches its attractor and a specific cost function is minimized (Buscema et al. 2006).

2.7   JNet: The Functional Scheme and Equations

The JNet is an ACM system developed for image analysis (edge extraction and segmentation). Its functional representation is showed in the flow chart (Fig. 2.23):

The scheme represents the iterative process based on the evolution of three main quantities: the minimal units U (which represent the image and its dynamical changing), the connections W (which represent the dynamic link between the units), and the state S of the system (which, combined with the connections W allows the whole system to converge).

We describe in detail the quantities involved in each step of the process (the superscript [n] on the quantities indicates the nth-step):

- *The Input*: This is a gray scale digital image. We consider an eight-bit gray scale image with dimensions W $\times$ H.
- *The set U of minimal units*: There is a minimal unit u_x for each pixel of the source image. They represent the nodes of this ANN. Every minimal unit u_x has a position $x = (x_1, x_2)$ with $x_1 = 1, ..., W$, $x_2 = 1, ..., H$ and an intensity value $u_x^{[n]}$. At the beginning each $u_x^{[n]}$ assumes the value of brightness of the pixel of the original image normalized in the range $[-1 + \alpha, 1 + \alpha]$, where $\alpha \in [0, 1]$.
- *The set W of connections*: For each pair of minimal units u_x and u_z we define the oriented connections $w_{x,z}^{[n]}$ and $w_{z,x}^{[n]}$. They depend on the positions of the minimal units $x = (x_1, x_2)$, $z = (z_1, z_2)$, $u_x^{[n]}$. At the beginning, each $w_{i,j}^{[0]}$ is equal and close to 0.

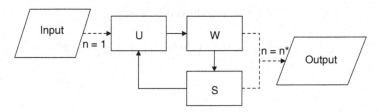

Fig. 2.23 JNet functional scheme

- *The activation state S:* There is a quantity $S_x^{[n]}$ for each minimal unit u_x (and therefore for each pixel of the source image). It is derived from the connections $w_{x,z}^{[n]}$.

The iterative process is based on the sequential update of the quantities W and U:

- *The update of the connections W* depends on the set U and the set W itself.
  For each position $i = (i_1, i_2)$ we consider the Moore neighborhood N(i) (i.e. the set of eight positions surrounding the position i) and the relative quantities $u_i^{[n]}$ and $w_{i,j}^{[n]}$:

$$D_i^{[n]} = \sum_{j \in N(i)} (u_j^{[n]} - w_{i,j}^{[n]}), \ J_i = \tanh(D_i) \tag{2.19}$$

$$\Delta w_{i,j}^{[n]} = -(u_i^{[n]} \cdot J_i) \cdot (-2 \cdot J_i) \cdot (1 - J_i^2) \cdot (u_j^{[n]} - w_{i,j}^{[n]}) , \ j \in N(i) \tag{2.20}$$

$$w_{i,j}^{[n+1]} = w_{i,j}^{[n]} + \Delta w_{i,j}^{[n]} \tag{2.21}$$

- *The calculation of the activation state S* depends on the set U and the set W.

$$MinW^{[n]} = \min_{i,j} \left(w_{i,j}^{[n]} \right) \tag{2.22}$$

$$MaxW^{[n]} = \max_{i,j} \left(w_{i,j}^{[n]} \right) \tag{2.23}$$

$$\begin{aligned} Scale_{Out}^{[n]} &= \frac{2}{MaxW^{[n]} - MinW^{[n]}} \\ Offset_{Out}^{[n]} &= -\frac{MaxW^{[n]} + MinW^{[n]}}{MaxW^{[n]} - MinW^{[n]}} \end{aligned} \tag{2.24}$$

$$AvW_i^{[n]} = \left| Scale_{Out}^{[n]} \cdot \frac{\sum_j^N w_{i,j}^{[n]}}{N} + Offset_{Out}^{[n]} \right| \tag{2.25}$$

$$\Delta S_i^{[n]} = -\tanh(AvW_i^{[n]} + u_j^{[n]}) \tag{2.26}$$

$$S_i^{[n]} = AvW_i^{[n]} + \Delta S_i^{[n]} = AvW_i^{[n]} - \tanh(AvW_i^{[n]} + u_j^{[n]}) \tag{2.27}$$

- *The update of the minimal units U depends on the set S and the set U itself.*

$$\varphi_i^{[n]} = LCoeff \cdot u_i^{[n]} \cdot \sum_j^N \left(1 - \Delta S_j^{[n]^2}\right) \text{ where } LCoeff \in [0,1] \tag{2.28}$$

$$\psi_i^{[n]} = \sum_j^N \tanh(\varphi_j^{[n]}) \tag{2.29}$$

We have alternative ways to combine the quantities $\varphi_i^{[n]}$ and $\psi_i^{[n]}$:

$$Version\ 1\ (sum) : \delta u_i^{[n]} = \varphi_i^{[n]} + \psi_i^{[n]} \tag{2.30a}$$

$$Version\ 2\ (product) : \delta u_i^{[n]} = \varphi_i^{[n]} \cdot \psi_i^{[n]} \tag{2.30b}$$

The choice between the two versions depends on the particular application. For the purpose of this study we have selected the first version [sum].

$$u_i^{[n+1]} = u_i^{[n]} + \delta u_i^{[n]} \tag{2.31}$$

The stop criterion of this cyclic evolution is connected to the stabilization of the connection's values $w_{x,z}^{[n]}$ (and consequently of the quantities $u_x^{[n]}$ and $S_x^{[n]}$). More precisely, we define the energy of the J-Net system, $E^{[n]}$, as the result of the addition of the changes of the connection's values relating the whole image to each processing cycle, according to the following equation:

$$E^{[n]} = \sum_{i \in X} \sum_{j \in N(i)} \left(\Delta w_{i,j}^{[n]}\right)^2 \tag{2.32}$$

where X is the set of pixels in the source image.

The evolution of the J-Net system determines a reduction of the system energy when the processing cycles increase:

$$\lim_{n \to \infty} E^{[n]} = 0. \tag{2.33}$$

This means that the stop criterion can be fixed in the following manner:

$$Stop\ when\ E^{[n]} < E_{Threshold} \ \Rightarrow \ n = n*, E^{[n*]} = E* \tag{2.34}$$

The energy of the system will be minimal at the end of the evolution:

$$E^* = \min\{E^{[n]}\} \ , \ n \in [1, ..., n*]. \tag{2.35}$$

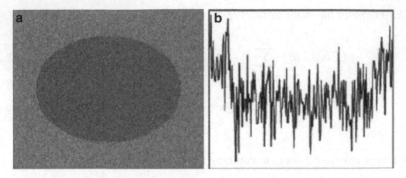

Fig. 2.24 (**a, b**) Noisy ellipse (*left*) and corresponding horizontal cross-section (*right*) taken at the center of the image. The size of the ellipse image is 253 × 189 pixels

Fig. 2.25 J-Net segmentation

Finally the output of the system is a gray scale digital image by which we can consider two different outputs:

- *Version 1 (states)*: the output is obtained by the state ($Out_i^{[n]} = \left| S_i^{[n]} \right|$).
- *Version 2 (weights)*: the output is derived from the weights ($Out_i^{[n]} = AvW_i^{[n]}$).

The choice between the two versions for output depends on the particular application. For the purpose of this study we have selected the first version (states).

2.7.1 JNet: Examples of Application

The J-Net Algorithm has two versions: Sum J-Net (2.30a) and Product J-Net (2.30b). Both the J-Net Algorithms, with a suitable defined *alpha* parameter, have been shown to work as an excellent adaptive filter for edge detection and segmentation (see Figs. 2.24, 2.25 and 2.26).

Fig. 2.26 J-Net edges
detection

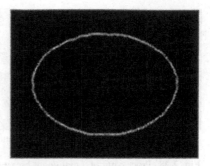

Fig. 2.27 Sum J-Net after
150 cycles

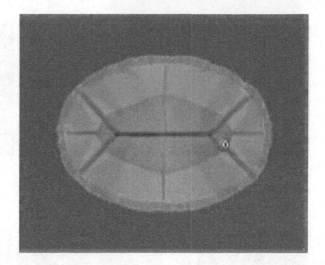

Sum J-Net after five cycles generates these results:

At the end of its processing Sum J-Net produces a special skeleton of the image, pointing the exact position of the two foci of the ellipsoid (see Fig. 2.27). The algorithm defines the skeleton of any image through propagation of waves from the edges of the figure up to its local centers. The destructive interferences of these waves define the skeleton figure.

The capability of Sum J-Net to find the edges, define the contour, segment and outline the skeleton of any image is a specific feature of this algorithm. Sum J-Net, consequently, is able to also define the hidden contour of an image and, using its wave's propagation process, make visible the hidden skeleton of the assigned image. Here are two examples of a breast spiculated carcinoma in X-ray (see Figs. 2.28 and 2.29).

The Sum J-Net capacity to detect a contour that is invisible to the radiologist is not dependent upon the physical generation of the image. Figure 2.30 is an example in which J-Net detects a hidden stenosis in an image of a poplitea arterial generated by a subtractive digital angiography machine, using a special contrast media (Buscema et al. 2008).

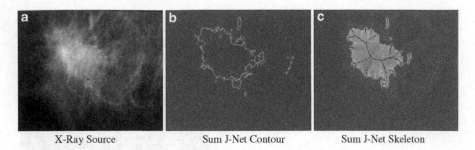

Fig. 2.28 Processing of a malignant breast mass

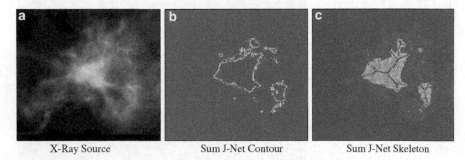

Fig. 2.29 Processing of a spiculated breast mass

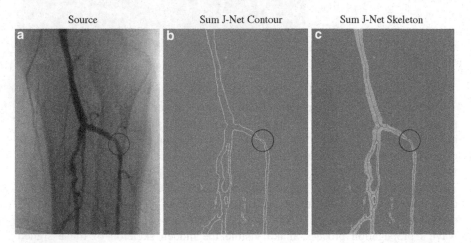

Fig. 2.30 (a, b, c) The hidden stenosis of a Popliteal Artery (Subtracting Digital Angiography with Contrast Media)

The Product J-Net Algorithm (2.12b) is particularly effective in detecting the multi segmentation analysis in which many areas are included in each other (see Fig. 2.31):

The main feature of the J-Net System is its ability to process the same image with different values of an *alpha parameter*. This kind of process permits us to generate a

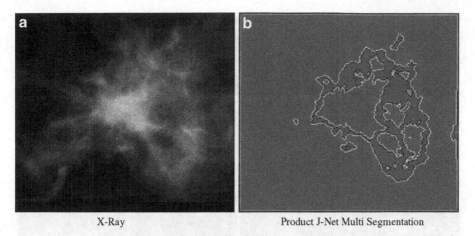

<table>
<tr><td>X-Ray</td><td>Product J-Net Multi Segmentation</td></tr>
</table>

Fig. 2.31 (a, b) Product J-Net algorithm, breast X-ray – Spiculed Carcinoma

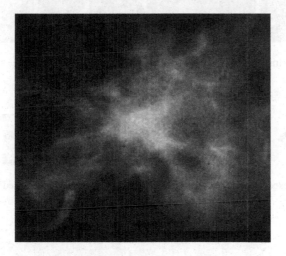

Fig. 2.32 Breast X-ray, particular of spiculed mass

different projection of the same image from the most evident contour to the one with the most light. The alpha parameter codes each different contour according to a different pseudo-frequency given each different projection. Low alpha values appear to correlate with some kind of low pseudo-frequency while high alpha values seem to be associated with some kind of high pseudo-frequency. We are currently unable to demonstrate this, but we have the impression, experimentally supported, that J-Net, though the modulation of this alpha parameter, transforms the light intensity of a pixel into a kind of pseudo-frequency. The real application we illustrate below shows a further experimental support for this hypothesis.

Figures 2.32 and 2.33 show the different contours of the same image using the Sum J-Net System with different tunings of the Alpha Parameter.

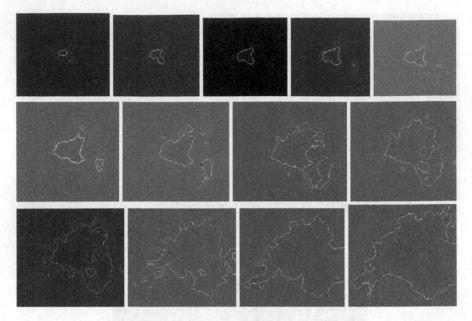

Fig. 2.33 J-Net with different Alpha Parameter, from alpha = −0.6 to alpha = +0.6 (shift = 0.1)

Fig. 2.34 J-Net expansion (Alpha increment) of a *benign* tumor ROI generated by ACF

Fig. 2.35 J-Net expansion (Alpha increment) of a *malignant* tumor ROI generated by ACF

Our hypothesis is that each different alpha parameter generates a new image whose information belongs to a past, or to a future, evolution of the original object from which the source image is derived given the original object is a living system evolving naturally over time. Figures 2.34 and 2.35 show an example of J-Net processing of a benign and of a malignant tumor from the artificial images generated by ACF (see above Figs. 2.21 and 2.22).

2.8   From J-Net to Histograms

This step is a simple way of coding the information generated by the J-Net Algorithm. Each lesion is represented by five J-Net ROIs and each J-Net ROI has a size of 101×101 pixels coded into 256 tones of grey. We have coded each J-Net ROI into a vector of 256 bins in which each bin is an integer number representing the number of times that each grey tone is present in the J-Net ROI (Fig. 2.36):

Because each lesion is represented by five J-Net ROIs, the union of the five histograms will code each lesion with a vector of 1,280 components (Fig. 2.37).

2.9   Population Algorithm for Multidimensional Scaling

The *Population* algorithm was conceived by *Massini* at the *Semeion Research Centre of Sciences of Communication* in 2006 (Massini et al. 2010). This algorithm fits into the theoretical framework for *Multi Dimensional Scaling*. The purpose of *Population* is to compress N records of a M dimensional space (*Source Space*) into a sub-space of P dimensions (*Projective Space*), where $P << M$, retaining as much information as possible about the relationship that exists between the original N records. *Population* is an iterative algorithm and it is based only on a calculation of *local fitness*. This fitness is considered optimal when the individual differences

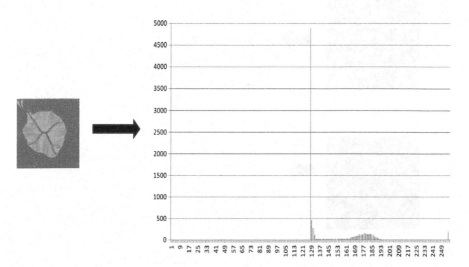

Fig. 2.36 Transformation of a J-Net ROI into a vector of 256 tones of grey

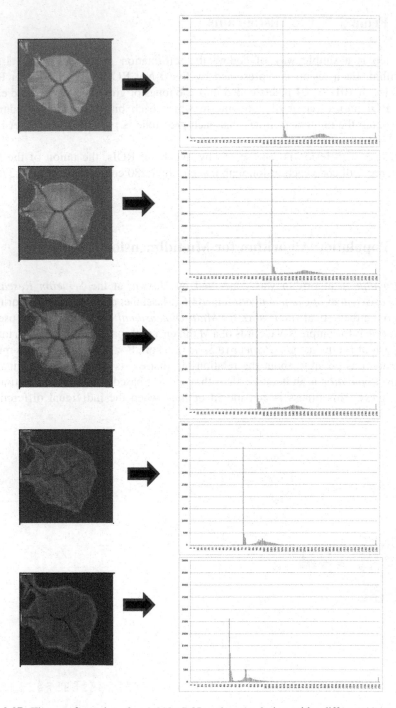

Fig. 2.37 The transformation of each J-Net ROI on the same lesion, with a different Alpha, into a vector of 1280 components

between the source matrix of distances and the projected matrix of distances are close to zero. Consequently, *Population* does not need to calculate its global fitness at each iteration. For this reason the convergence of *Population* is very fast compared to other algorithms of *Multi Dimensional Scaling*, such as that of Sammon (1969). It is therefore especially useful for the design of very large databases (above 100 K records).

At the beginning, the values of the new vectors of the projective space are generated randomly, and each iterative cycle consists of the following steps:

- Random selection of two records V_i and V_j;
- Calculation of the distance vector RD_{ij} between the vectors V_i and V_j of the *Source Space* and the distance $MD_{i,j}$ between the vectors $V'_i V'_j$ of the Projective *Space*.

$$RD_{i,j} = \sqrt{\sum_{k}^{M} \left(v_{i,k} - v_{j,k}\right)^2} \; ; \tag{2.36}$$

$$MD_{i,j} = \sqrt{\sum_{k}^{P} \left(v'_{i,k} - v'_{j,k}\right)^2} \; . \tag{2.37}$$

- Calculation of the error (ER) between the distances RD_{ij} and MD_{ij}

$$ER_{i,j} = RD_{i,j} - MD_{i,j}. \tag{2.38}$$

- Correction of each vector V' so that the difference between RD_{ij} and MD_{ij} is reduced.

The correction factor Δ is added or subtracted to minimize the difference ER_{ij} between the distances RD_{ij} and MD_{ij}. All the Δ factors are calculated in proportion to the error ER_{ij}. In practice, when $MD_{i,j} > RD_{i,j}$, the Δ factor is calculated by (2.39a), otherwise (2.39b) is used (obviously, when $MD_{i,j} = RD_{i,j}$ no correction is applied):

$$\Delta_{i,j,k} = \left(v'_{i,k} - v'_{j,k}\right) \cdot \left(1 - \frac{RD_{i,j}}{MD_{ij}}\right); RD_{i,j} < MD_{i,j}; \tag{2.39a}$$

$$\Delta_{i,j,k} = -\left(v'_{i,k} - v'_{j,k}\right) \cdot \left(1 - \frac{MD_{i,j}}{RD_{i,j}}\right); RD_{i,j} \geq MD_{i,j}. \tag{2.39b}$$

Finally, the correction factor, scaled by a constant (*alpha*), is added to each projected vector $V'_{i,k}$:

$$V'_{i,k_{(n+1)}} = V'_{i,k_{(n)}} + \sum_j \Delta_{i,j,k} \cdot \alpha$$

(2.40)

$$\text{typically} : \alpha = 0.1$$

The iterative process of *Population* converges toward a classic minimization cost function:

$$Energy = Min\left\{\frac{1}{2}\sum_{i\neq j}^{N}\sum_{j}^{N}\left(RD_{i,j} - MD_{i,j}\right)^2\right\}$$

(2.41)

Regarding the performances on large datasets we emphasize the fact that for *Population* the saving of a distance matrix is not necessary because each correction is made in real time (online). The greater advantage is obviously the speed of the elaboration system because it does not need to calculate the *fitness, stress* or other global functions of cost in order to minimize the error. The end of the elaboration can be determined, therefore, using two different criteria:

• Sampling of the cost function,
• Average of the error percentage.

The sampling of the cost function is executed randomly in the sample, every *n* cycles, and the error is calculated only on the matrix of their mutual distances. The iteration is interrupted when the error becomes stable. The average of the percentage of the error is calculated on the average and variance of the corrections made during each *n* cycles. The elaboration is interrupted when the average of the error becomes stable.

The Population program has demonstrated that it possesses a much higher resolution quality for the multi-dimensional scaling problem (Massini et al. 2010). The potential for this algorithm is considerable:

1. Speed enhancement;
2. Efficiency improvement;
3. Simplicity of the algorithm;
4. Freedom from having to calculate a specific cost function;
5. The possibility of analyzing a dataset of great dimension;
6. The possibility of dynamically introducing new records into the dataset during the program run;
7. The possibility of choosing the dimensions of Projected Space.

Figure 2.38 shows an application of this algorithm on the known "IRIS" dataset. Figure 2.39 shows another application of Population on a very large dataset.

Algorithm 1

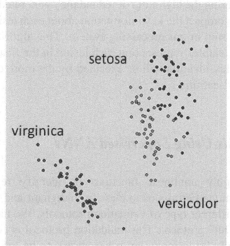

Fitness 0.9289 / Stress 0.9920

Fig. 2.38 Elaboration with population relative to two different cost functions based on data of 150 records (*Iris*) for 4 variables

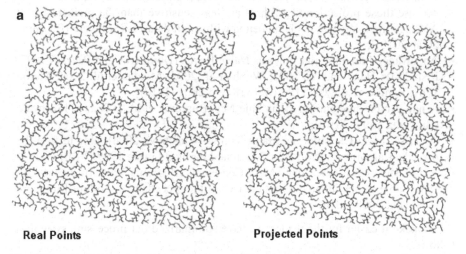

Fig. 2.39 (a, b) Elaboration of population on a 5,000 point dataset with random coordinates XY within a minimum and a maximum value. On the left the MST is calculated on the original coordinates and on the right the MST is calculated on the coordinates population identified

2.10   Application of Population Algorithm to Medical Data

We have used the Population algorithm to reduce the dimensionality of the data generated by the union of histograms of the five J-Net ROIs for each lesion. We have always chosen the minimum dimension of the projection space so as to be able to maintain an error of projection smaller than 0.01. This strategy has shown to be effective in filtering the high ratio of redundancy generated by the histograms, and consequently to compact the key information about each lesion. The Population algorithm is the last step in our processing system. This algorithm has the task of producing the set of features representing each lesion in the final dataset. After this step, the definitive classification will be executed by the more classical supervised ANNs and Machine Learning.

2.10.1   Diagnosis Using Supervised ANNs

ANNs compute highly nonlinear functions to identify relationships among variables. This technique permits us to classify malignant and benign lesions in a "blind" way using different type of validation protocols. We have avoided the use of the "Leave One Out" protocol. This validation protocol is convenient for small datasets (we have only 90 lesions), but it has shown to be too unstable. We have instead chosen the more robust K-Fold Cross Validation ($K = 5$) as validation protocol and have used and compared different Learning Machine Algorithms:

- an enhanced version of LVQ ANN (Adaptive Vector Quantization (Kohonen 1995–2001; Kosko 1992; Neuralware 1995; Buscema and Catzola 2010), because these projection algorithms are less sensitive than the gradient based ANNs to the overtraining phenomenon with small datasets [Software (Buscema 2008–2010)];
- An advanced learning machines as Naive Bayes (Domingos and Pazzani 1997; Rish 2001; Hand and Yu 2001; Webb et al. 2005; Mozina et al. 2004; Maron 1961; Rennie et al. 2003) [Software (Rapid 2001–2010)];
- A new ANN, gradient based, as Sine Net (Buscema et al. 2006a, b) [Software (Buscema 1999–2010)];
- A classic Meta Classifier, as Major Voting (Kuncheva 2004), able to fuse the precedent algorithms [Software (Machine Learning Group 1999–2010)];
- A new Meta Classifier, named Meta-Consensus (Buscema 1998; Buscema et al. 2010), able to maximize the best choices of the precedent algorithms [Software (Buscema 2008–2010)].

To make it easier for the reader, we have summarized our processing system as follows:

1. Proposed Processing System, composed of three basic steps: Fusion, Expansion, and Squashing (FES):

   (a) ACF Fusion: All the slices of each lesion are fused by the ACF system;
   (b) J-Net: J-Net (Weights–Union) system re-writes each fused lesion into many images;
   (c) Grey Histograms Vector: codification of J-Net ROIs into a vector representing the tones of grey of each lesion;

Fig. 2.40 Fusion and J-Net segmentation of a benign lesion

Fig. 2.41 Fusion and J-Net segmentation of a malignant lesion

    (d) Population: the dimensionality of each matrix lesion is squashed into a small number of features, maintaining the key information of the assigned matrix;

    (e) Classification: using the new features as input vector of each lesion, supervised ANNs try to blindly classify the new lesions dataset.

        We have also analyzed the same data, using more classical approaches. In this way it is possible to understand the advantages of every processing step proposed in comparison with other less expensive and standard analyses.

2. Naive Analysis (NA):

    (a) ACF Fusion: All the slices of each lesion are fused by ACF system;

    (b) Population: Multidimensional Scaling of the input vector of 101×101 original pixels into a more compact number of input features;

    (c) Final supervised classification.

3. Classic Analysis (CA):

    (a) ACF Fusion: All the slices of each lesion are fused by ACF system;

    (b) Binning of each ROI: each ROI is coded into a 256 input vector. Each vector component is coded with the number of presences of any grey tone in each lesion;

    (c) Population: Multidimensional Scaling of the input vector of 256 variables into a more compact number of input features

    (d) Final supervised classification.

4. Object Oriented Analysis (OOA):

    (a) ACF Fusion: All the slices of each lesion are fused by the ACF system;

    (b) Segmentation of each ROI into two areas: the lesion area and the background. We have used J-Net Union-State to make this segmentation (see Figs. 2.40 and 2.41);

    (c) Calculation of the area and perimeter of the lesion, and calculation of the homologous circle having the same area of the lesion and their ratio;

    (d) Final supervised classification using the precedent features.

2.11 Results

We have processed the 90 lesions in two different ways:

1. An enhanced LVQ ANN (AVQ (Buscema and Catzola 2010)) was applied for the final supervised classification to the output of our system (FES) and to the output of the concurrent systems (NA, CA, and OOA) in five independent blind tests (K-Fold CV, K = 5) for each processing system.
2. Three different learning machines and two meta classifiers were applied only to the outputs of our system, using always a K-Fold CV (K = 5) validation protocol.

2.11.1 Systems Comparison

In Tables 2.1, 2.2, 2.3 and 2.4 we show the initial results. The J-Net system, with the alpha processing, marks the difference from the other systems (in each table "A.Mean" stands for the *arithmetic* mean and "W.Mean" stands for the *weighted* mean).

Table 2.1 Naïve analysis, results with K-Fold CV

Naive analysis (Na)					
K-Fold(NA)	Malignant (%)	Benign (%)	A.Mean (%)	W.Mean (%)	Error
FF_AVQ1_G(1)	90.00	100.00	95.00	94.74	1
FF_AVQ1_G(2)	88.89	100.00	94.44	94.12	1
FF_AVQ1_G(3)	77.78	88.89	83.33	83.33	3
FF_AVQ1_G(4)	88.89	77.78	83.33	83.33	3
FF_AVQ1_G(5)	100.00	77.78	88.89	88.89	2
AVQ (Mean)	89.11	88.89	89.00	88.88	2

Table 2.2 Classic analysis, results with K-Fold C

Classic analysis (CA)					
K-Fold (CA)	Malignant (%)	Benign (%)	A.Mean (%)	W.Mean (%)	Error
FF_AVQ1_G(1)	100.00	88.89	94.44	94.74	1
FF_AVQ1_G(2)	88.89	100.00	94.44	94.12	1
FF_AVQ1_G(3)	77.78	100.00	88.89	88.89	2
FF_AVQ1_G(4)	66.67	100.00	83.33	83.33	3
FF_AVQ1_G(5)	100.00	77.78	88.89	88.89	2
AVQ (Mean)	86.67	93.33	90.00	89.99	1.8

Table 2.3 Object oriented analysis, results with K-Fold CV fusion, expansion and squashing (FES)

Object oriented analysis (OOA)					
K-Fold(OOA)	Malignant (%)	Benign (%)	A.Mean (%)	W.Mean (%)	Error
FF_AVQ1_G(1)	80.00	77.78	78.89	78.95	4
FF_AVQ1_G(2)	88.89	100.00	94.44	94.12	1
FF_AVQ1_G(3)	88.89	88.89	88.89	88.89	2
FF_AVQ1_G(4)	55.56	100.00	77.78	77.78	4
FF_AVQ1_G(5)	77.78	88.89	83.33	83.33	3
AVQ(Mean)	78.22	91.11	84.67	84.61	2.8

Table 2.4 FES system, results with K-Fold CV

K-Fold(FES)	Malignant (%)	Benign (%)	A.Mean (%)	W.Mean (%)	Error
FF_AVQ1_G(1)	100.00	100.00	100.00	100.00	0
FF_AVQ1_G(2)	88.89	100.00	94.44	94.12	1
FF_AVQ1_G(3)	77.78	100.00	88.89	88.89	2
FF_AVQ1_G(4)	100.00	100.00	100.00	100.00	0
FF_AVQ1_G(5)	100.00	77.78	88.89	88.89	2
AVQ (Mean)	93.33	95.56	94.44	94.38	1.0

Table 2.5 FES System using AVQ

K-Fold(FES)	Malignant (%)	Benign (%)	A.Mean (%)	W.Mean (%)	Error
FF_AVQ1_G(1)	100.00	100.00	100.00	100.00	0
FF_AVQ1_G(2)	88.89	100.00	94.44	94.12	1
FF_AVQ1_G(3)	77.78	100.00	88.89	88.89	2
FF_AVQ1_G(4)	100.00	100.00	100.00	100.00	0
FF_AVQ1_G(5)	100.00	77.78	88.89	88.89	2
AVQ	93.33	95.56	94.44	94.38	1.0

2.11.2 Results Using Meta Classifiers Optimization

In this new experiment we have analyzed the output of our system (FES) using three different machine learning algorithms:

- An enhanced Vector Quantization (AVQ);
- A Naïve Bayes Classifier (Naïve Bayes)[3];
- A Sine Net ANN (SN).

In Tables 2.5, 2.6 and 2.7 we show the results of three independent K-Fold CV applied to each algorithm:

AVQ and Naïve Bayes outperform SN ANN. AVQ shows a good sensibility and a very good specificity. Naïve Bayes shows an excellent sensitivity. However, SN has an interesting performance, quite orthogonal to the other machines learning. The diversity of these performances is the right prerequisite to combine all these three classifiers into a meta-classifier, to take the best of their specific capabilities. To implement this fusion we have selected two Meta-Classifiers:

- Major Voting, a classic and a simple algorithm able to combine different classifiers (Kuncheva 2004), and
- Meta-Consensus Net, a new meta-classifier, recently presented to the scientific community (Buscema et al. 2010).

---

[3] All the experimentations with Naive Bayes algorithm were executed by Massimiliano Marciano (software engineer at CSI Research and Innovation, via Cesare Pavese 305, Rome, Italy), using Rapid Miner ver. 5.0.010.

Table 2.6 FES system using Naïve Bayes

K-Fold(FES)	Malignant (%)	Benign (%)	A.Mean (%)	W.Mean (%)	Error
NaiveBayes(1)	90.00	87.50	88.75	88.89	2
NaiveBayes(2)	100.00	85.71	92.86	93.75	1
NaiveBayes(3)	100.00	87.50	93.75	94.12	1
NaiveBayes(4)	100.00	87.50	93.75	94.12	1
NaiveBayes(5)	100.00	87.50	93.75	94.12	1
NaiveBayes	98.00	87.14	92.57	93.00	1.2

Table 2.7 FES system using Sine Net

K-Fold(FES)	Malignant (%)	Benign (%)	A.Mean (%)	W.Mean (%)	Error
FF_Sn(1)	70.00	100.00	85.00	84.21	3
FF_Sn(2)	88.89	62.50	75.69	76.47	4
FF_Sn(3)	77.78	77.78	77.78	77.78	4
FF_Sn(4)	77.78	100.00	88.89	88.89	2
FF_Sn(5)	77.78	88.89	83.33	83.33	3
SN	78.45	85.83	82.14	82.14	3.2

Table 2.8 FES system using Major Voting

K-Fold(FES)	Malignant(%)	Benign (%)	A.Mean (%)	W.Mean (%)	Error
MajorVote(1)	90.00	100.00	95.00	94.74	1
MajorVote(2)	100.00	100.00	100.00	100.00	0
MajorVote(3)	88.89	100.00	94.44	94.44	1
MajorVote(4)	100.00	100.00	100.00	100.00	0
MajorVote(5)	100.00	77.78	88.89	88.89	2
MajorVote	95.78	95.56	95.67	95.61	0.8

Table 2.9 FES system using Meta-Consensus Net

K-Fold(FES)	Malignant (%)	Benign (%)	A.Mean (%)	W.Mean (%)	Error
Meta-Consensus(1)	100.00	100.00	100.00	100.00	0
Meta-Consensus(2)	100.00	100.00	100.00	100.00	0
Meta-Consensus(3)	100.00	88.89	94.44	94.44	1
Meta-Consensus(4)	100.00	100.00	100.00	100.00	0
Meta-Consensus(5)	100.00	88.89	94.44	94.44	1
Meta-Consensus	100.00	95.56	97.78	97.78	0.4

Tables 2.8 and 2.9 show the K-Fold CV results when each of the two Meta-Classifiers is applied to the outputs of the three basic classifiers:

Meta-Consensus outperforms all the precedent results and enhances the quality of the information extraction process executed through the J-Net algorithm.

2.11.3   Discussion

The first description of a lung nodule characterization CAD is ascribed to Kawata et al. (1998). These authors in 1998 described a CAD scheme for distinguishing between benign and malignant nodules. Their scheme first utilized a deformable surface model to extract nodule regions based on an initial surface placed within a nodule. They then extracted three features from the segmented nodule regions, i.e., the attenuation, shape index, and curvedness value. The shape index measures whether a surface is convex or concave, and the curvedness reflects the degree of curvature on a surface. The histograms of the attenuation, shape index, and curvedness value for pixels within a nodule region were obtained, and the scale of each histogram was employed as a feature. A Fisher linear classifier was trained to provide a score for distinction between benign and malignant nodules. A leave-one-out method was utilized for evaluating the performance level of the CAD scheme based on a total of 62 patients, including 35 with malignant nodules and 27 with benign nodules. Each patient was scanned three times at three specific time points: before the injection of a contrast agent and two and four minutes after the start of contrast administration. The Az values (where A is the area under the curve and z is the original score), when all three features were used for the three time points, were 0.91 ± 0.04, 0.99 ± 0.01, and 1.0, respectively. The sensitivity and specificity values were 94% and 74% for the CT images scanned before the injection of contrast agent, 100% and 89% two minutes after contrast administration, and 100% and 100% four minutes after contrast administration.

McNitt-Gary et al. (1999) also developed a CAD scheme for the distinction between benign and malignant nodules. Their database contained 35 patients, including 19 with malignant and 16 with benign nodules. All of the patients had at least one volumetric scan and may have had up to four scans imaged 45, 90, 180, and 360 seconds after the injection of contrast agent. Their scheme first employed a semi-automated procedure to segment nodule regions. From a seed point identified by a user of the scheme, a region growing algorithm with user-adjustable upper and lower thresholds was utilized to create a nodule region. The segmented nodule region was reviewed, edited, and approved by one of three thoracic radiologists in their team. Each segmented nodule region was then further partitioned into two regions: one containing only a solid portion and the other containing only a ground-glass portion. For each of the two regions of every nodule, 31 features were calculated, including 12 attenuation features, five size features, four shape features, and 10 contrast enhancement features. Feature selection was accomplished by a stepwise model selection search by the Akaike Information Criterion so that the extent of over fitting was reduced during the subsequent classification step. For three feature sets including 31 features extracted from the solid portion, 31 features from the ground-glass portion, and 62 features from both portions, the feature selection method selected 6, 6, and 5 features, respectively. It seemed that features extracted from the ground-glass portion were not very effective for distinguishing between benign and malignant nodules, regardless whether they were used alone or combined with features from the solid portion. Three classifiers, including linear and quadratic discriminant analysis as well as logistic regression, were employed to distinguish between benign and malignant nodules. A leave-one-out method was utilized for evaluating the performance level of the CAD scheme. It appeared that the logistic regression classifier provided the highest performance level, and its Az value (Area under the ROC curve) for distinction between benign and malignant nodules was 0.92.

Aoyama described a CAD scheme for nodule characterization (Aoyama 2003) by use of a dynamic programming technique. Forty-one and 15 image features based on 2D sectional data and 3D volumetric data, respectively, were determined from quantitative analysis of the nodule outline and of pixel values. A stepwise feature selection method extracted eight features which were input to a linear classifier for distinguishing between benign and malignant nodules. A leave-one-out testing method was employed to evaluate the performance of this CAD scheme. A total of 244 patients, including 61 with malignant and 183 with benign nodules, constituted the study population. The area under the ROC curve was employed to measure the performance level of this CAD scheme. The CAD scheme yielded an overall ROC/AUC of 0.937 (0.919 for nodules with pure ground-glass opacity, 0.852 for nodules with mixed ground-glass opacity, and 0.957 for solid nodules) for distinction between the 61 malignant and 183 benign lung nodules. These are, in our view, the best reference papers on this subject.

In comparison with the existing literature concerning CAD systems applied to lung nodules characterization, our approach has two major differences: the first is the use of a sequential set of adaptive systems rather than a single system. This has to do with the trivial concept that complex problems require complex approaches to be solved. By using an assembly of algorithms based of different mathematics it is easier to capture the intrinsic complexity of the information responsible for the discrimination among malignant and benign lesions.

The second difference relies in the validation protocols employed to circumstantiate scientifically the predictive power of the algorithms. At variance with existing literature we have not employed the leave-one-out approach but rather a K fold cross-validation protocol. The choice is justified by the fact that the leave-one-out (LOO) approach is not appropriate for small data sets like ours being too unstable (LOO over fits data and its results are over pessimistic (Kunceva 2004), while K-fold cross validation is more robust to sampling variability).

The discriminating capacity among malignant and benign lesions obtained in this study is one, if not the best thus far, described in the literature. This was not completely expected but it is also not a shocking surprise. We think these results constitute a strong rationale for further study to be performed on a larger set of patients, hopefully enrolled by different clinics, to guarantee that a generalization of the results is confirmed outside the original center. We also need to compare the performance of the multi-system with the performance of the radiologist in a blind way. The use of improved ANNs as classifiers will not solve all known problems of lung MDCT. Manually obtained morphological features and the measurement of signal intensities inside manually placed ROIs still needs be to well standardized.

One reason for the good results achieved during this study may be the fact that only two radiological experts evaluated these subjective features in close cooperation. A combination of more radiologists will surely increase the inter-observer variability of extracted features. This could lead to an increased number of errors by the ANNs. To avoid this problem, the next step must be the introduction of automatic feature extraction algorithms.

An early diagnosis represents the main goal to achieve when dealing with patients with small lung nodules; this would allow a curative treatment for malignant tumors to begin at early stages, when healing of the patient is still possible. Our system seems to be able to enhance nodule morphologic features not visible to the human eye that might be a reflection of the modifications related to the growth and invasion of malignant cells in the closest periphery and in the inner regions of a nodule. The results achieved in our investigation look extremely promising, and our approach may open new perspectives in the clinical and radiological management of patients with small lung nodules.

References

Aoyama M, Li Q, Kasuragawa S, MacMahon H, Doi K (2002) Automated computerized scheme for distinction between benign and malignant solitary lung nodules on chest images. Med Phys 29:701–708

Aoyama M, Li Q, Katsuragawa S, Li F, Sone S, Doi K (2003a) Computerized scheme for determination of the likelihood measure of malignancy for lung nodules on low-dose CT images. Med Phys 30(3):387–394

Aoyama M, Li Q, Katsuragawa S, Li F, Sone S, Doi K (2003b) Computerized scheme for determination of the likelihood measure of malignancy for pulmonary nodules on low-dose CT images. Med Phys 30:387–394

Blum RS, Liu Z (2006) Multi-sensor image fusion and its applications. Taylor & Francis, London

Buscema M (1998) Meta-Net: the theory of independent judges. In: Buscema M (ed) Substance use & misuse. Marcel Dekker Inc., New York, vol 33(2), pp 439–461

Buscema M and Catzola L (2010) Adaptive vector quantization systems: classic LVQ and new AVQ. Semeion Institute, Rome, IT, Mimeo (in Italian)

Buscema M et al. (2006) Sistemi ACM e imaging diagnostico. Le immagini mediche come matrici attive di connessioni [ACM systems and diagnostic imaging. Medical images as active connections matrices, in Italian] Springer-Verlag, Italy

Buscema M, Breda M and Terzi S (2006a) A feed forward sine based neural network for functional approximation of a waste incinerator emissions. Proceedings of the 8th WSEAS international conference on automatic control, modeling and simulation, Prague

Buscema M, Breda M, Terzi S (2006c) Using sinusoidal modulated weights improve feed-forward neural network performances in classification and functional approximation problems. WSEAS Trans Inf Sci Appl 3(5):885–893

Buscema M, Catzola L, Grossi E (2008) Images as active connection matrixes: the J-Net system. IC-MED Int J Intell Comput Med Sci 2(1):27–53

Buscema M, Terzi S, Tastle W (2010) Meta-consensus: a new meta classifier. In: NAFIPS 2010, 12–14 July 2010, Toronto, Canada, 978-1-4244-7858-6/10 ©2010 IEEE

Diederich S, Wormanns D, Heindel W (2003) Lung cancer screening with low-dose CT. Eur Radiol 45(1):2–7

Diederich S, Wormanns D, Semik M (2002) Screening for early lung cancer with low-dose spiral CT: prevalence in 817 asymptomatic smokers. Radiology 222:773–781

Domingos P, Pazzani M (1997) On the optimality of the simple Bayesian classifier under zero–one loss. Mach Learn 29:103–137

Fischbach F, Knollmann F, Griesshaber V, Freund T, Akkol E, Felix R (2003) Detection of lung nodules by multislice computed tomography: improved detection rate with reduced slice thickness. Eur Radiol 13:2378–2383

Goldin J, Brown M, Petkovska I (2008) Computer-aided diagnosis in lung nodule assessment. J Thorac Imaging 23(2):97–104

Greenlee RT, Murray T, Bolden S, Wingo PA (2000) Cancer statistics, 2000. CA Cancer J Clin 50:7–33

Hand DJ, Yu K (2001) Idiot's Bayes – not so stupid after all? Int Stat Rev 69(3):385–399, ISSN 0306–7734

Henschke CI, Yankelevitz DF (2008) CT screening for lung cancer: update 2007. Oncologist 13:65–78

Henschke CI, Yankelevitz DF, Libby D et al (2002) CT screening for lung cancer: the first ten years. Cancer J 8:S47–S54

Kawata Y, Niki N, Ohmatsu H, Kakinuma R, Eguchi K, Kaneko M, Moriyama N (1998) Quantitative surface characterization of pulmonary nodules based on thin-section CT images. IEEE Trans Nucl Sci 45:2132–2138

Kohonen T (1995–2001) Self-organizing map, 3rd edn. Springer, Berlin

Kosko B (1992) Neural networks and fuzzy systems. Prentice Hall, New Jersey, pp 39–261

Kunceva LI (2004) Combining pattern classifier. Wiley Interscience, Hoboken, NJ

Kuncheva LI (2004) Combining pattern classifiers: methods and algorithms. Wiley, Hoboken, NJ

Li F, Sone S, Abe H, MacMahon H, Armato AG, Kunio D (2002) Lung cancer missed at low-dose helical CT screening in a general population: comparison of clinical, histopathologic, and imaging findings. Radiology 225:673–683

Li Q (2007) Recent progress in computer-aided diagnosis of lung nodules on thin-section CT. Comput Med Imaging Graph 31(4–5):248–57

Maron ME (1961) Automatic indexing: an experimental inquiry. J ACM (JACM) 8(3):404–417

Massini G, Terzi S and Buscema M (2010) A new method of multidimensional scaling. In: NAFIPS 2010, 12–14 July 2010, Toronto, Canada, 978-1-4244-7858-6/10 ©2010 IEEE

McNitt-Gary MF, Hart EM, Wyckoff N, Sayre JW, Goldin JG, Aberle DR (1999) A pattern classification approach to characterizing solitary pulmonary nodules imaged on high resolution CT: preliminary results. Med Phys 26:880–888

Mozina M, Demsar J, Kattan M, Zupan B (2004) Nomograms for visualization of naive Bayesian classifier. In: Proceedings of PKDD-2004, pp 337–348

Neuralware (1995) Neural computing. A technology for professional II/PLUS and neural networks explorer. NeuralWare Inc., Pittsburg, PA, pp 227–233

Rennie J, Shih L, Teevan J, Tackling K (2003) The poor assumptions of naive Bayes classifiers. In: Proceedings of the twentieth international conference on machine learning (ICML), Washington, DC

Rish I (2001) An empirical study of the naive Bayes classifier. IJCAI 2001 Workshop on empirical methods in artificial intelligence, Seattle, USA

Sammon JW (1969) A non linear mapping for data structure analysis. IDEE Trans Comput C-18 (5):401–409

Shiraishi J, Abe H, Engelmann R, Aoyama M, MacMahon H, Doi K (2003) Computer-aided diagnosis to distinguish benign from malignant solitary pulmonary nodules on radiographs: ROC analysis of radiologists' performance – initial experience. Radiology 227:469–474

Shiraishi J, Li Q, Suzuki K, Li F, Engelmann R, Doi K (2006) Computer-aided diagnostic scheme for the detection of lung nodules on chest radiographs: Localized search method based on anatomical classification. Med Phys 2642:2653

Swensen SJ, Jett JR, Hartman TE, Midthun DE, Sloan JA, Sykes AM, Aughenbaugh GL, Clemens
    MA (2003) Lung cancer screening with CT: Mayo clinic experience. Radiology 226:756–761
Webb GI, Boughton J, Wang Z (2005) Not so naive Bayes: aggregating one-dependence
    estimators. Mach Learn 58(1):5–24

Software

Buscema M (1999–2010) Supervised ANNs and organisms, ver. 16.5. Semeion software #12,
    Rome, Italy
Buscema M (2003–2010) Active connections matrix, ver. 13.5. Semeion software #30, Rome Italy
Buscema M (2008–2010) Meta-Nets, ver. 8.0. Semeion software #44, Rome, Italy
Buscema M (2010) ACF: image fusion, ver.3.5. Semeion software #55, 2010, Rome, Italy
Machine Learning Group (1999–2010) WEKA. Waikato environment for knowledge analysis, ver.
    3.6.3, the University of Waikato, Hamilton, New Zealand
Massini G (2007–2009) Population: an fast algorithm for multi-dimensional scaling, ver.3.0.
    Semeion software #42, Rome, Italy
Rapid I (2001–2010) Rapid Miner, ver. 5.0.010, Dortmund, Germany [http://www.rapidminer.com]

Patents

Buscema M (2003) European patent "An algorithm for recognizing relationships between data of a
    database and a method for image pattern recognition based on the said algorithm" (Active
    Connection Matrix –ACM). Owner: Bracco, Semeion. Application no. 03425559.6, deposited
    August 22, 2003
Buscema M (2004) International patent "An algorithm for recognizing relationships between data
    of a database and a method for image pattern recognition based on the said algorithm" (Active
    Connection Matrix – ACM). Owner: Bracco, Semeion. Application n. PCT/EP2004/05182,
    deposited: August 18, 2004
Buscema M (2007) European patent "Active Connection Matrix J-Net. An algorithm for the
    processing of images able to highlight the information hidden in the pixels". Owner: Bracco,
    Semeion. Application n. 07425419.4, deposited: July 6, 2007

Chapter 3
Population Algorithm: A New Method of Multi-Dimensional Scaling

Giulia Massini, Stefano Terzi, and Massimo Buscema

3.1 Introduction and Motivation

The Population algorithm (hereafter referred to as Population) has a place in the theoretical framework of Multi Dimensional Scaling. Reducing the dimensionality of a dataset is a frequent problem in the analysis of data and is remarkable important, in particular, in the field of exploratory analysis. Population provides an opportunity to compress N records of a M-dimensional space (which we call the Source Space) in a subspace of Q dimensions (called the Projected Space), where $Q << M$, maintaining the greatest possible number of existing relations contained in the original N records. Population is an iterative algorithm based on the calculation of a local fitness, the distance between two points that is considered optimal when the single differences between the matrix of the distances of the Source Space and the matrix of the distances of Projected Space are near zero.

This particular characteristic of Population, the ability to converge on a solution without calculating the global fitness, determines the speed with which it finds a solution minimizing the global error compared to other algorithms of Multi Dimensional Scaling, such as that of Sammon (1969). It is therefore particularly useful for elaborations of datasets of great dimensionality; we consider a dataset of great dimensionality to be or the order of magnitude of some 100K+ records.

G. Massini (✉) • S. Terzi
Semeion Research Center of Sciences of Communication, Via Sersale 117, Rome, Italy
e-mail: g.massini@semeion.it

M. Buscema
Semeion Research Center of Sciences of Communication, Via Sersale 117, Rome, Italy

Department of Mathematical and Statistical Sciences, CCMB, University of Colorado, Denver, Colorado, USA

W.J. Tastle (ed.), *Data Mining Applications Using Artificial Adaptive Systems*, DOI 10.1007/978-1-4614-4223-3_3, © Springer Science+Business Media New York 2013

The Sammon method consists of the minimization of the Stress Function

$$E = \frac{1}{\sum\limits_{i<j} \left[d_{ij}^*\right]} \sum\limits_{i<j} \frac{\left[d_{ij}^* - d_{ij}\right]^2}{d_{ij}^*}$$

Where:

d_{ij}^* is the distance of Source Space;
d_{ij} is the distance of Projected Space.

The more recent global approaches (Isomap – Tenenbaum et al. 2000), and local approaches (Locally Linear Embedding (Roweis and Saul 2000), Laplacian Eigenmaps (Belkin and Niyogi 2003)) have been addressed towards the search for algorithms that give various embedded results, such as the modification of a cost function. Contrary to these recent approaches that have concentrated on the definition of new modalities of projection to conserve different characteristics from the matrix of the distances in the original space, we have chosen to take, as our initial reference, the approach of Sammon (1969).

3.2 The Algorithm

In Population the algorithm is based on a calculation of some local fitness considered optimal when the single differences are calculated by the matrix of the distances between the vectors of dimension M and the matrix of the distances between the vectors of dimension Q, where (Q << M), are near zero.

The algorithm can be expressed as follows:

1. Randomly select two records V_i and V_j.
2. Calculate the vectorial distance RD_{ij} between the V_i and V_j vectors of the Source Space and of the MD_{ij} distance between V_i^I and V_j^I vectors of Projected Space.

$$RD_{ij} = \sqrt{\sum_{k}^{M} \left(v_{i,k} - v_{j,k}\right)^2} \qquad (3.1)$$

$$MD_{ij} = \sqrt{\sum_{k}^{Q} \left(v_{i,k}' - v_{j,k}'\right)^2} \qquad (3.2)$$

3. Calculate the Error (ER) between the RD_{ij} and MD_{ij} distances

$$ER_{ij} = RD_{ij} - MD_{ij} \qquad (3.3)$$

4. Correction of the V_i^I vector is necessary so that the difference between RD_{ij} and MDij is reduced.
5. The correction factor (Δ) is added or subtracted to minimize the difference ER_{ij} between the RD_{ij} and MD_{ij} distances. All Δ factors are calculated in proportion to

ER_{ij} error. When $MD_{ij} > RD_{ij}$, the Δ factor will be calculated with (3.4), otherwise, when $MD_{ij} < RD_{ij}$, the factor is calculated with (3.5). When $MD_{i,j} = RD_{i,j}$ there is no correction.

$$\Delta_{ijk} = \left(v'_{ik} - v'_{jk}\right) \cdot \left(1 - \frac{RD_{ij}}{MD_{ij}}\right); RD_{ij} < MD_{ij} \tag{3.4}$$

$$\Delta_{ijk} = -\left(v'_{ik} - v'_{jk}\right) \cdot \left(1 - \frac{MD_{ij}}{RD_{ij}}\right); RD_{ij} > MD_{ij} \tag{3.5}$$

However, since Population is not bound to a specific Cost Function, it is possible to define the Δ of correction with respect to the considered objective. Another way to calculate the correction factor (Buscema, (2009) Personal communication) that has given optimal results is the following:

$$\Delta_{ijk} = \left(v'_{ik} - v'_{jk_i}\right) \cdot \frac{\left(MD_{ij} - RD_{ij}\right)^2}{RD_{ij}{}^2}; RD_{ij} < MD_{ij} \tag{3.6}$$

$$\Delta_{ijk} = -\left(v'_{ik} - v'_{jk_i}\right) \cdot \frac{\left(MD_{ij} - RD_{ij}\right)^2}{RD_{ij}{}^2}; RD_{ij} \geq MD_{ij} \tag{3.7}$$

At the end, the correction factor, scaled by the constant α, is added to the projected vector $V_{ik}{}^I$:

$$V'_{ik(t+1)} = V'_{ik(t)} + \sum_j \Delta_{ijk} \cdot \alpha$$

$$where \ \alpha = 0.1 \tag{3.8}$$

The iterative process of Population aims, therefore, to minimize the Cost Function:

$$Energy = Min\left\{\frac{1}{2}\sum_{i \neq j}^{N}\sum_{j}^{N}\left(RD_{ij} - MD_{ij}\right)^2\right\} \tag{3.9}$$

The following pseudocode shows the operation of the algorithm:

```
PROCEDURE PopulationRun;
BEGIN
  REPEAT
    inc (cycles);
    random extraction of 2 records (Vi, Vj);
    calculation of vectorial distance RD(Vi, Vj);
    calculation of vectorial distance MD(V^I i, V^I j);
    IF (dg<>dv) THEN
      Correction_of_vectorial_distance_MD(V^I i, V^I j);
  UNTIL stop := true;
END;
```

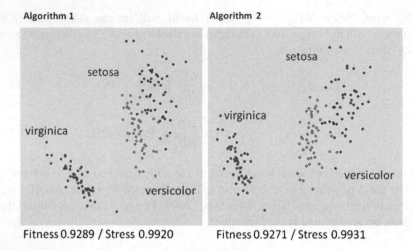

Fitness 0.9289 / Stress 0.9920   Fitness 0.9271 / Stress 0.9931

Fig. 3.1 Elaboration with Population relative to two different cost functions based on data of 150 records (*Iris*) for four variables

We have evaluated the results of Multi Dimensional Scaling algorithms using two cost functions:

1. Stress Function

$$\text{Stress} = \frac{1}{\sum_{i \neq j} \text{RD}_{ij}} \cdot \sum_{i=j}^{N-1} \sum_{j=i+1}^{N} \frac{\left(\text{RD}_{ij} - \text{MD}_{ij}\right)^2}{\text{RD}_{ij}} \tag{3.10}$$

2. Fitness Function

$$\text{Fitness} = 1 - \left(\frac{2}{C} \cdot \sum_{i=j}^{N-1} \sum_{j=i+1}^{N} \frac{\left(\text{RD}_{ij} - \text{MD}_{ij}\right)^2}{\text{RD}_{ij}} \right) \tag{3.11}$$

Where $C = N^2 - N$

In Fig. 3.1 we use Population to show the result of an analysis of two different functions on the calculation of the Δ of correction (Algorithm 1 with (3.4 and 3.5) or (Algorithm 2 with 3.6 and 3.7) based on the data from 150 records related to three varieties of Iris flowers (Virginica, Versicolor, Setosa) regarding four characteristics (Sepali Length, Sepali Width, Petals Length, Petals Width). The results of the elaborations are slightly different with respect to the two fitness/stress functions of cost and are shown as Algorithms 1 and 2. With Algorithm 1 the Fitness is improved while with Algorithm 2, Stress is improved.

3.3   Results

For experiments of comparison between the Population algorithm and others, we develop one program using Delphi (Massini 2007–2009) and another in Matlab (SOM Toolbox). In the implementation of Matlab it was not necessary to include the Sammon algorithm, for there was already an implementation in the SOM Toolbox of Matlab.

The experiments are based on two kinds of verifications: (a) that the abilities of convergence of the algorithm on the stress function and of the fitness function are analogous to those of Sammon's classic algorithm and (b) that the speed is higher in the Population approach.

A first test was done on a limited database of 8 records and 3 variables, relative to coordinates x, y, z of a 3D cube to be projected onto a 2D space. The results are shown in figs. 3.2 and 3.3.

With this first experiment we observe that the Population algorithm has identified a solution comparable to that obtained with the Sammon algorithm. Then we tested both algorithms on a set of four datasets: Segment, Letter and Satim taken from the UCI repository (Asuncion and Newman 2007) and Defects belonging to Semeion repository (see Table 3.1).

This comparison shows evidence of a small difference of about 0.01 in terms of an ability to minimize Stress in favour of the original Sammon algorithm, due probably to the fact that the inherent function of cost in the Population algorithm is quite different from that of the classic function of Stress.

The differences are, however, so small that we are unable to invalidate the use of the Population algorithm as an alternative to Sammon. The comparison of the speed of the algorithms was not based exclusively on temporal terms, since such an approach could have suffered from different optimizations resulting from the writing of the code. Therefore we have chosen to calculate the number of steps used by the two algorithms and also to use a second measure that is tied to the number of base operations executed.

Fig. 3.2 Result using the Sammon algorithm on eight coordinates x, y, z of a 3D cube projected in 2D space

Stress 0.9379; Fitness 0.8111

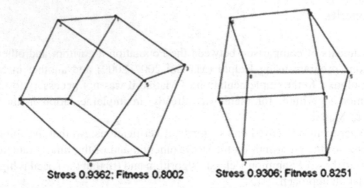

Stress 0.9362; Fitness 0.8002   Stress 0.9306; Fitness 0.8251

Fig. 3.3 Result using the Population (algorithm 1) on eight coordinates x, y, z of a 3D cube projected onto a 2D space. The two figures have been obtained by saving best run (SAVE BEST) of Stress or, vice versa, the best value of Fitness

Table 3.1 Comparison of Population vs Sammon results

			Population				Sammon		
	Record	Var	Algorithm	Stress	Fitness	Cycles	Stress	Fitness	Epochs
Defects	389	27	2	0.93	0.76	75,333	0.93	0.77	248.1
Letters	200	16	1	0.92	0.78	639,833	0.93	0.78	262.7
Satimage	322	36	1	0.98	0.88	355,588	0.98	0.88	237
Segments	231	18	1	0.96	0.84	742,068	0.97	0.84	406.6

Table 3.2 Comparison of Population vs Sammon results from computer time consuming point of view

			Population			Sammon			Report
	R	V	Stress	Fitness	Operations	Stress	Fitness	Operations	Pop/Sam
Defects	389	27	0.93	0.76	451,998	0.93	0.77	337,016,063	745.61
Letters	200	16	0.92	0.78	3,839,598	0.93	0.78	94,099,140	24.51
Satimage	322	36	0.98	0.88	2,133,528	0.98	0.88	220,471,146	103.34
Segments	231	18	0.96	0.84	4,452,408	0.97	0.84	194,423,922	43.67

For Population, every cycle is based on the calculation operation of the existing difference between two distances and on the single correction (~6 operations), while every epoch of the algorithm of Sammon is based on ~9 operations that have to be multiplied by n*(n−1) each time the Stress factor is calculated on the total of the comparisons.

In Table 3.2 the cycles/epochs have been replaced with an appraisal of the number of operations executed.

Table 3.2 shows that Population completes a considerable smaller number of operations; however, in temporal terms the higher speed of Population does not produce great advantages with small databases, rather, when the databases are very large the difference becomes much more meaningful.

Fig. 3.4 Results of the Population algorithm on a DB of 5,000 points, stopped at about 3,100,000 cycles (≈18,600,000 operations)

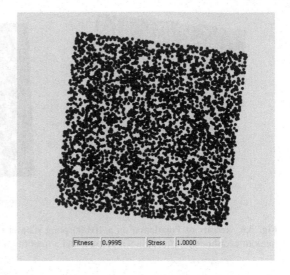

Fig. 3.5 Results of the Population algorithm on a 5,000 point dataset with random coordinates XY within a minimum and a maximum value. On the left the MST is calculated on the original coordinates and on the right the MST is calculated on the coordinates Population identified

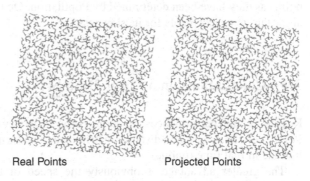

Real Points   Projected Points

For these simulation experiments we have used an artificial dataset that is representative of real bi-dimensional maps with a huge number of points. Convergence is considered met when the value of residual stress drops below 10^{-4}.

In the text below we show the result of three experiments carried out on datasets composed of 5,000, 10,000, and 50,000 points with random coordinates XY between a minimal and maximal value. The result is a square form composed by points of increasing density. In all cases Population reached the value of 1.000 for Stress (Fig. 3.4).

Compared to a dataset of 5,000 points we show the Minimum Spanning Tree (MST) obtained by combining the points on the plan with respect to their own original coordinates and the coordinates obtained with Population. As Fig. 3.5 shows, the two MST's do not permit the crossing of branches.

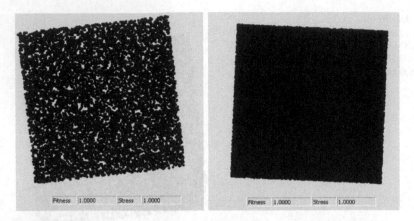

Fig. 3.6 Results of Population on a 10,000 point dataset (*left*) and 50,000 points (*right*) with random coordinates XY within a minimum and a maximum value

Figure 3.6 shows the images of the other two datasets of 10,000 and 50,000 points as they have been determined by Population. On these datasets the Sammon algorithm is unacceptable for its slowness.

3.4 Calculations on a Large Data Set

Regarding the performances on large datasets we emphasize the fact that for Population the saving of a distance matrix is not necessary because each correction it made in real time (online).

The greater advantage is obviously the speed of the computational system because it does not need to calculate the fitness, stress or other global functions of cost in order to minimize the error.

The end of the elaboration can be decided, therefore, using two different criteria:

• Sampling of the cost function,
• Average of the error percentage.

The sampling of the cost function is executed randomly in the sample, every n cycles, and the error is calculated only on the matrix of mutual distances. The iteration is interrupted when the error becomes stable. The average of the percentage of the error is calculated on the average and variance of the corrections made each n cycles. The calculation is interrupted when the average of the error becomes stable. Moreover, a strong possibility exists to *use Population to pre-process data that then could be refined by another program to obtain some desired result*; this application of Population should not be underestimated. Population is already used to pre-process data in combination with the PST (Buscema 1999–2009; Buscema and Terzi 2006) program that implements a specific genetic algorithm

Table 3.3   156 variables of the macro-array dataset

1	B101N	27	B63T	53	B98T	79	B111T	105	B2-T	131	B2-L
2	B46N	28	B66T	54	B100T	80	B113T	106	B31-T	132	B31-L
3	B49N	29	B67T	55	B101T	81	B81T	107	B32-T	133	B32-L
4	B65N	30	B68T	56	B102T	82	B88T	108	B33-T	134	B33-L
5	B81N	31	B69T	57	B104T	83	B92T	109	B3-T	135	B3-L
6	B83N	32	B71T	58	B105T	84	B94T	110	B4-T	136	B4-L
7	B94N	33	B72T	59	B106T	85	B95T	111	B5-T	137	B5-L
8	B96N	34	B74T	60	B107T	86	B96T	112	B6-T	138	B6-L
9	B10-Nrep	35	B75T	61	B112T	87	B99T	113	B8-T	139	B8-L
10	B11-Nrep	36	B76T	62	B115T	88	B7-T	114	B9-T	140	B9-L
11	B12-N	37	B77T	63	B78T	89	B10-T	115	B10-L	141	B25-Lneg
12	B16-N	38	B53T	64	B79T	90	B11-T	116	B11-L	142	B30-Lneg
13	B18-N	39	B58T	65	B80T	91	B12-T	117	B12-L	143	B64L
14	B2-N	40	B60T	66	B82T	92	B13-T	118	B13-L	144	B7-Lneg
15	B3-N	41	B70T	67	B83T	93	B14-T	119	B14-L	145	B19-Met
16	B9-N	42	B42T	68	B84T	94	B15-T	120	B15-L	146	B20-Met
17	B110T	43	B43T	69	B86T	95	B16-T	121	B16-L	147	B21-Met
18	B114T.	44	B44T	70	B87T	96	B17-T	122	B17-L	148	B22-Met
19	B52T	45	B45T	71	B89T	97	B18-T	123	B18-L	149	B34-Met
20	B54T	46	B46T	72	B90T	98	B1-T	124	B1-L	150	B35-Met
21	B55T	47	B47T	73	B91T.	99	B23-T	125	B23-L	151	B36-Met
22	B56T	48	B48T	74	B93T	100	B24-T	126	B24-L	152	B37-Met
23	B57T	49	B49T	75	B97T	101	B26-T	127	B26-L	153	B38-Met
24	B59T	50	B50T	76	B103T.	102	B27-T	128	B27-L	154	B39-Met
25	B61T	51	B51T	77	B108T.	103	B28-T	129	B28-L	155	B40-Met
26	B62T	52	B73T	78	B109T	104	B29-T	130	B29-L	156	B41-Met

(genD-Buscema 2004) able to optimally minimize the fitness function in a space projected in 2/3D.

It is useful to also show an application of Population on a large dataset: the dataset we have considered is a Macro -Array composed of 22,283 records and 156 variables (Table 3.3). The results of the calculation of all 22,283 records are shown in Fig. 3.7. This figure shows four different close ups of the final map provided by Population.

3.5   Discussion of the Dynamics of a Data Set

Because of the nature of the Population algorithm, it is not necessary that the data set be complete at the moment of use. New data can be dynamically introduced into the system. The procedure used for appending to a data set in which the sequence of records is temporal, follows:

During the elaboration a new record, $V_{t(n + 1)}$, is positioned into the Projected Space according to the rule of the triangulation related to the last three records previously introduced in t_n, t_{n-1}, t_{n-2}. The system continues to analyze the dataset that has now increased to include the new record.

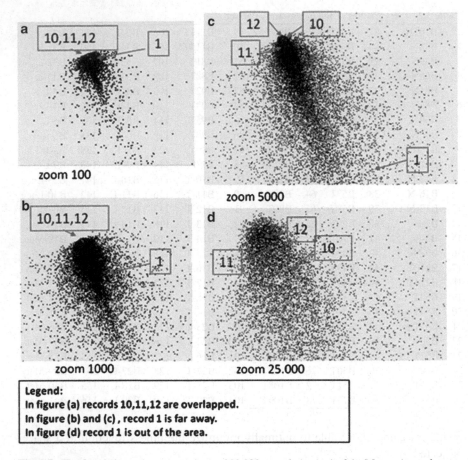

Legend:
In figure (a) records 10,11,12 are overlapped.
In figure (b) and (c) , record 1 is far away.
In figure (d) record 1 is out of the area.

Fig. 3.7 The figure shows four zoom views of 22,283 records (*points*) of the Macro-Array dataset at the end of Population processing

The map of the Projected Space therefore approaches its shape based on the first records of the data set because there is no need to wait for a complete dataset to be provided.

An example follows. A dataset composed of five nano-sensors is based on the collection of evidence related to an introduction of poison into a room. A determination is made on a sequence of 10,559 signals for every sensor, and each record is composed of five values. The data have been standardized in table format. The sequence of the signal data for the five values is shown in Fig. 3.8.

Using Dynamic Population to assess the first 696 records results in the 2D map of Fig. 3.9a. The entire final calculation on all the 10,559 records is shown in Fig. 3.9b, c. The configuration shows two "tails": on the right the initial records and on the left the final ones, since the signal tends to return to the initial values it had at rest conditions. The elaboration was conducted by introducing one record at a time using the triangulation method.

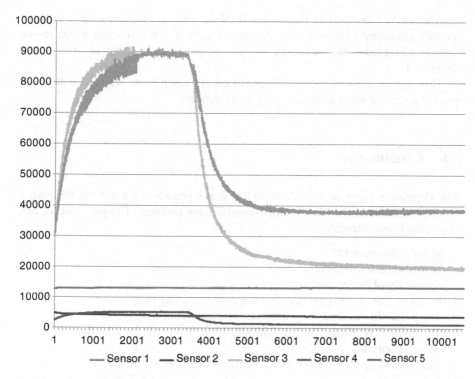

Fig. 3.8 Visualization on a graph of the signals of five nano-sensors in presence of poison in the air

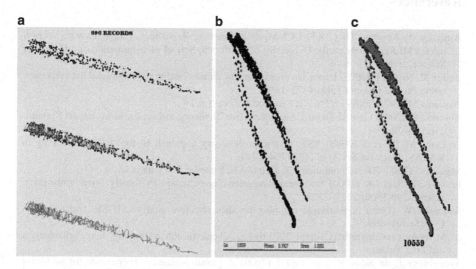

Fig. 3.9 Elaboration with Dynamic Population on the signals of five nano-sensors in presence of poison in the air. In (**a**) the 2d configuration and the map of the first 696 records. In (**b**) the 2d configuration of all the 10,559 records and in (**c**) the map

After every introduction, Population fixed the distances between the introduced records extracting randomly, two records at a time, for a number equal to the introduced records. At the end of this procedure we have a fitness of 0.99 and stress = 1.

This method allows a result to be produced in a much faster way and with much greater precision when analyzing very large datasets.

3.6 Conclusions

The Population program has demonstrated that it possesses a much higher quality for the resolution of the Multi-Dimensional Scaling problem. The potential for this algorithm is considerable:

1. Speed enhancement;
2. Efficiency improvement;
3. Simplicity of the algorithm;
4. Freedom from having to calculate a specific cost function;
5. The possibility of analyzing a dataset of great dimension;
6. The possibility of dynamically introducing new records into the dataset during the program run;
7. The possibility of choosing the dimensions of Projected Space.

References

Asuncion A, Newman DJ (2007) UCI Machine Learning Repository. [http://www.ics.uci.edu/mlearn/MLRepository.html]. University of California, School of Information and Computer Science, Irvine

Belkin M, Niyogi P (2003) Laplacian eigenmaps for dimensionality reduction and data representation. Neural Comput 15(6):1373–1396

Buscema M (1999–2009) PST v.9.1. Semeion Software n. 11

Buscema M (2004) Genetic Doping Algorithm (GenD): theory and applications. Expert Systems. 21(2):63–79

Buscema M, Terzi S (2006) PST: a new evolutionary approach to topographic mapping in WSEAS. Trans Inf Sci Appl 3(9):1704–1710

Massini G (2007–2009) Population v.3, Delphi 6.0, Semeion Software n.42

Roweis ST, Saul LK (2000) Nonlinear dimensionality reduction by locally linear embedding. Science 290(5500):2323–2326

Sammon JW (1969) A nonlinear mapping for data structure analysis. IDEE Trans Comput C-18(5):401–409

SOM Toolbox is Copyright (C) 2000–2005 by Esa Alhoniemi, Johan Himberg, Juha Parhankangas and Juha Vesanto

Tenenbaum J, de Silva V, Langford J (2000) A global geometric framework for nonlinear dimensionality reduction. Science 290(5500):2319–2323

Chapter 4
Semantics of Point Spaces Through the Topological Weighted Centroid and Other Mathematical Quantities: Theory and Applications

Massimo Buscema, Marco Breda, Enzo Grossi, Luigi Catzola, and Pier Luigi Sacco

Part A: Syntax of Physical Space: Theory

4.1 The Conceptual Context

4.1.1 Introduction

The *spatial dimension* is often a key feature to understand the structure of a phenomenon. In some cases, this is relatively obvious, as phenomena themselves are basically defined in spatial terms as in the case of, for example, diffusion processes. In some other cases, however, such dimension is not obviously relevant. For example, think of a comparative analysis of different socio-economic systems in which geographical coordinates are not necessarily part of the data base. There is a vast array of alternative, statistically based approaches that have been developed to deal with the spatial character of phenomena, each approach based on different fields such as physics, biology, economics and geography, and many more. But an aspect that has been somewhat overlooked so far is that of the semantic dimension of space, that is to say the interpreting of the topological or metric dimension of space as conveying an *intrinsic* meaning that may have a substantial bearing on the

M. Buscema (✉)
Semeion Research Center of Sciences of Communication, Via Sersale 117, Rome, Italy

Department of Mathematical and Statistical Sciences, CCMB, University of Colorado, Denver, Colorado, USA
e-mail: m.buscema@semeion.it

M. Breda • L. Catzola • P.L. Sacco
Semeion Research Center of Sciences of Communication, Via Sersale 117, Rome, Italy

E. Grossi
Bracco Research Division, Via Folli, 50, 20134 Milan, Italy

W.J. Tastle (ed.), *Data Mining Applications Using Artificial Adaptive Systems*,
DOI 10.1007/978-1-4614-4223-3_4, © Springer Science+Business Media New York 2013

interpretation of the underlying phenomena. Thus, whether or not we are considering a physical space or some sort of abstract, representational space, we must consider the possibility that the spatial dimension may carry *relational* information as to why certain entities 'stay together' in a given environment; this may add substantially to our understanding of the respective phenomena. It is important to stress the somewhat original meaning that we give to the term 'relational' in this context: We are claiming that, to make any kind of significant sense of a certain phenomenon, we must consider its variables as negotiating their position within the state space according to a semantic defined in terms of *relative proximity*. In other words, the physical entities produced by a given phenomenon situate themselves in the state space as if they were *aware,* to some degree, of their relative position and would adjust to each other in appropriate ways, as if they were abiding by a sort of *implicit grammar* of the phenomenon. We can unscramble and interpret such negotiation by suitably re-mapping the space in such a way as to give proximity to its most *expressive* meaning. Insofar as the semantic aspects of the phenomenon relate to the spatial characteristics only, we are basically reasoning in terms of a *syntax* of space, i.e., the *internal* rules by which spatial features combine appropriately to generate proper structures of meaning. This is the most basic level of a relational spatial analysis. If, in addition, the phenomenon presents some characteristic dimensions of a non-spatial nature, we can speak of a full *semantic*. In this case, we will have to develop a more articulated approach that will be presented in the second part of the chapter.

In this chapter, we present a number of new mathematical quantities that are particularly useful at capturing such relational dimensions and thus to allow for a rigorous analysis of the semantics of space. Specifically, such quantities are practical and relatively easy to use in cognitively accessible spaces – namely, two- or three-dimensional ones. More indirectly, they can also be used to analyze the semantics of points defined in higher dimensional space. In this case, a multidimensional scaling algorithm should be previously applied in order to obtain a projection of the high-dimensional *source space* onto a two- or three-dimensional *target space*. For the target space semantics to be representative of the original source space, it is necessary that the scaling algorithm be capable of minimizing the distortion of the hyper-distances once they are projected from the source to the target space. The smaller the distortion error, the more the target conveys a semantic that is congruent to the original one.

Since R-record of a V-variables dataset can always be seen as a set of R points in a V-dimensional space, the introduced quantities can be used to describe some semantic aspects of the dataset that are related to the relative position of the R records in their V-dimensional space. By the same token, one may transpose the reasoning and regard the same quantities as illustrative of some semantic aspect of the relative positions of the V variables in their R-dimensional space, provided that such transposition is computationally feasible.

All the proposed quantities will be defined considering a set of K points, called *entities*, in a two-dimensional space – the extension to the three dimensional case is straightforward. As we will see, the proposed mathematical quantities are points, curves or scalar fields. They are listed below and then defined in the specific sections:

- Topological Weighted Centroid (TWC)[1];
- Self Topological Weighted Centroid (STWC);
- Proximity Scalar Field;
- Gradient of the Scalar Field;
- Relative Topological Weighted Centroid (TWC$_i$);
- Paths from the Arithmetic Centroid to entities;
- Paths between entities;
- Scalar Field of the trajectories.

These quantities imply that each entity in the set of K points has the same features apart from its position within the N-dimensional space. But we can, additionally, introduce the possibility that each entity has different features other than its position, thereby causing more complex interactions with the others within the N-dimensional ambient space. To deal with this further complication, we will apply a new algorithm, named the Auto Contractive Map, to evaluate the relationships among the K non-homogeneous entities while taking their specific qualitative features into account. Furthermore, we will propose a new method for combining our derived relationships among non-homogeneous entities with the pre-existing geographical information.

Overall, we have thus developed a conceptually innovative methodology to deal with space semantics that may prove particularly interesting and effective in tackling particularly complex problems and even some kinds of problems that are commonly believed to be intractable according to the currently available toolbox of methods and methodologies. To fully illustrate the scope and power of such techniques, we apply them, in the second part of the paper, to a variety of different problems taken from various disciplines whose heterogeneity makes a clear case for the 'universality' of our approach.

4.1.2 Location Theory

In the scientific literature the problem of the semantics of the geographic space is typically analyzed within the framework of Location Theory (Buscema et al. 2009a; Brantingham and Brantingham 1981, 1984; Levine 2004; O'Leary 2006, Buscema & Terzi, 2006, 2006a).

Location theory is concerned with one of the central issues in geography. This theory attempts to find an optimal location for any particular distribution of activities, population, or events over a region according to a specific criterion.

---

[1] The Topological Weighted Centroid and its equations were designed by M. Buscema in 2008 at Semeion.

The specific location problem we want to deal with can be simply defined in the following terms. Let's consider a distribution of K points in a N-dimensional space, $X_i = \{x_{i_1}, \ldots, x_{i_N}\}$, with $i = 1, \ldots,$ K, and typically $N = 2$ or $N = 3$, depending on whether we want to deal with a two or three dimensional space. Let these points be spatial positions representing locations where something meaningful happened related to an existing phenomenon under study. They could be, for instance, locations where deaths occurred in case of a disease outbreak, or crime scenes where serial offences took place. We want to define and calculate a point function $H : \Re^N \to \Re[0, \infty]$, or, more specifically, its extremal value:

$$Y^* = \arg\max_{Y \in \Re^N} \{H(X_1, \ldots, X_K, Y)\}$$

where H expresses the likelihood of finding the originating point(s) of the phenomenon. Fixing an appropriate scaling factor, A, the function $H(X_1, \ldots, X_K, Y)$ can be seen as a probabilistic function:

$$H(X_1, \ldots, X_K, Y) = A \cdot P(X_1, \ldots, X_K | Y)$$

and the given problem as a problem of maximization of the likelihood:

$$Y^* = \arg\max_{Y \in \Re^N} \{P(X_1, \ldots, X_K | Y)\}$$

The function's extremal value, for instance, could identify the origin of the outbreak or the serial offender's hiding place.

4.1.3 Benchmark Solutions

Several solutions can be put forward for the problem just described. The most relevant ones for the purposes of this paper are described as benchmark elements for the proposed algorithm. The simplest solution is based on the estimation of the central location from which travel distance and/or time is minimized, the so called *anchor point*:

$$Y^* = \arg\min_{Y \in \Re^N} \left\{ \sum_{i=1}^{K} D^N(Y, X_i) \right\}$$

Where $D^N(Y, X_i)$ is the distance between the anchor point Y and each of the given points X_i, based upon some given metric.

Whereas the anchor point minimizes some kind of distance, the *center of mass* minimizes the square of that (Euclidean) distance. We therefore have a second possible, very simple option to consider as a solution to the given problem:

$$Y^* = \underset{Y}{\arg\min} \left\{ \sum_{i=1}^{K} \sum_{r=1}^{N} (Y_r - X_{i_r})^2 \right\}$$

Once again, however, this point need not be of particular interest for a location theory with realistic purposes.

4.1.4   General Approach

In order to reconstruct the underlying law that generates the observed scatter of points, we have to take into account more information than just time and distance. Every agent (say, a predator or a virus) has a given quantity of energy to spend. Such energy is consumed during travel (such as a virus expansion or hunting). The energy consumption can determine the length of the hunt or the strength of the virus' expansion. Other random incidents, such as physical or psychological constraints, can also interrupt travel. We thus have to consider that the given scatter of points will typically be the result of some kind of resource (energy expenditure) optimization. Reasonable locations, then, must take this contextual information into account. In the case of a predator, for example, the optimal point is, as a rule, the location that is close to the majority of the hunting areas. In this situation, any failure in predation can be easily compensated by spending a little more energy in traveling to the next near hunting area. The same is true for a virus; the more people in a given area, the greater the virus' capacity to infect and replicate. In the case of human predators (serial criminals), a place not so distant from the predator's house, and yet close enough to the majority of assault places, is ideal. Thus, if we want to develop a usable location theory, we have to consider not only distances, but also the agent's energy minimization strategy and the geographical and contextual constraints of the territory.

We must then define an *operation point* as the point which minimizes suitably rescaled distances, as well as all of the relevant constraints whatever their nature (geographic, energetic, and so forth). In particular, a general enough model for $P(X_1, \ldots, X_K | Y)$ can be provided by considering:

$D^N(Y, X_i)$: distances between the operation point Y and each of the given points X_i according to a specific metric (Euclidean distance, Block distance, Travel distance, Travel time, etc);

$F\big(D^N(Y, X_i)\big)$: distance decay function;

$G(X_i)$: a function representing geographic factors and other context-specific features;

$E(X_i,Y)$: a function representing the energy distribution of the field in which the given (observed) points are located; and

$S(r_1, \ldots, r_K)$: a composition function.

On the basis of the above, we can write:

$$P(X_1, \ldots, X_K | Y) = S\left(\begin{array}{l} F\big(D^N(Y,X_1)\big) \cdot G(X_1) \cdot E(X_1,Y), \ldots \\ , F\big(D^N(Y,X_K)\big) \cdot G(X_K) \cdot E(X_K,Y) \end{array} \right)$$

The classic algorithms that are found in the current literature consider all of these components, but $G(X_i)$. We will show how our approach, the TWC (Topological Weighted Centroid) theory and its algorithms, with the support of the Auto Contractive Map Neural Network, is able to also take into account the $G(X_i)$ component, that is, the semantics of the physical space.

The specific geographic and non-geographic factors characterizing the space where each observed point is located represent a key point in the location theory. In most cases, location theory algorithms do not systematically deal with these features. This is the reason why location theory in its current state actually represents a way to analyze the *syntax* of the space. TWC algorithms with the Auto Contractive Map can, therefore, be regarded as the first systematic approach to the *semantics* of the physical space.

4.1.5 Probability Distribution Strategies

This class of algorithms focuses only on the metric of the space and the shape of the decay distance function, and uses the sum as a composition function without specific assumptions about other factors. Following this approach, the operational point is located in a region with a high "Hit Score":

$$Y^* = \underset{Y \in \Re^N}{\arg\max} \{P(X_1, \ldots, X_K | Y)\} = \underset{Y \in \Re^N}{\arg\max} \{S(X_1, \ldots, X_K, Y)\}$$

$$S(X_1, \ldots, X_K, Y) = \sum_{i=1}^{K} F\big(D^N(Y,X_i)\big)$$

More practically, these approaches tend to define the probability of each point – within the convex hull of the locations of the observations – to be the operational point. Consequently, the probability distribution strategy defines a search area whose points have the highest probability of being the operational point (Buscema et al. 2009a; Brantingham and Brantingham 1981, 1984; Levine 2004; O'Leary 2006; Rich and Shively 2004; Rossmo 2005).

In this chapter we consider the two well-known algorithms of this kind:

- The Rossmo Algorithm (Rossmo 1993, 1995, 1997, 2000, 2005); and
- The Canter Algorithm (Canter and Larkin 1993; Canter and Taggs 1975; Canter 2003, 1999; Canter et al. 2000).

In addition, we consider also three proprietary algorithms, two of which have already been presented in a previous paper (Buscema et al. 2009a):

- The Negative Exponential Summation Algorithm (NES); and
- The Likelihood with Variance Maximization Algorithm (LVM).

The third algorithm is presented here for the first time, after one year of testing:

- The Mexican Probability Algorithm (MexProb).

These five algorithms, together with the Anchor Point and the Center of Mass, will be compared to the TWC algorithms with some real problems for which the actual operational point is known.

4.1.6   The Rossmo Algorithm

The Rossmo Algorithm uses the block distance. It employs four free parameters (B, k, g, h), each of which has to be calibrated empirically according to the situation. This algorithm is specific to the identification of an operational point in a serial crime. We adapt its four parameters and metric and apply it in the quest to track an epidemic. The Rossmo equations are the following:

$$F\left(D^N(Y, X_i)\right) = F(d) = \begin{cases} \dfrac{k}{d^h} & if(d > B) \\ \dfrac{k \cdot B^{g-h}}{(2B-d)^g} & if(d \leq B) \end{cases}$$

The first term of the equation takes its inspiration from Newtonian gravity. The whole equation represents a formulation of the Brantingham and Brantingham (1981, 1984) search area model in which the offender's search behaviour is seen as following a distance decay function with decreased activity near the offender's home base. Rossmo has produced examples showing how the model can be applied to serial offenders (Rossmo 2000, 2005). For both the 'within buffer zone' (near to home base, controlled by the parameter B) and 'outside buffer zone' (far from home base) functions, the parameter k and the exponents h and g are empirically determined. Though he doesn't discuss how these are calculated, they are presumably estimated from a sample of known offenders' home-bases in cases for which the distance to each crime scene is known (e.g., through arrest records).

4.1.7   Canter Algorithm

Canter's group in Liverpool (Canter and Larkin 1993; Canter and Taggs 1975; Canter 1999, 2003; Canter et al. 2000) have suitably modified the distance decay function for offenders' choices of assault locations by using a negative exponential term instead of the inverse distance:

$$F\left(D^N(Y,X_i)\right) = F(d) = A \cdot e^{-\beta \cdot d}$$

Also the Canter algorithm was devised for the detection of serial offenders, and thus we have employed it with the same cautions used to adopt the Rossmo algorithm.

4.1.8   The NES Algorithm

The NES Algorithm (Buscema et al. 2009a) uses the buffer concept of the Rossmo model and the negative exponential of Canter, combining them in the following equations:

$$F\left(D^N(Y,X_i)\right) = F(d) = 1 - e^{-v(d)}$$

$$v(d) = \varphi \cdot e^{-d \cdot g} + (1 - \varphi) \cdot e^{-B \cdot (g-h)} \cdot e^{-(d-2B) \cdot g}$$

When h and g are fine-tuned appropriately (h = 0.05 and g = 0.01) and the NES Algorithm has proven to be very sensitive to the distribution of the observed sites.

4.1.9   The LVM Algorithm

The LVM Algorithm (Buscema et al. 2009a) is inspired by the O'Realy Bayesian model. The main difference is the cost function: We try to maximize the variance of the likelihood among all the candidate operation points, by means of an iterative process:

$$F\left(D^N(Y,X_i)\right) = F\left(D^N(Y,X_i)_{\alpha*}\right) = F(d_{\alpha*})$$

$$F(d_\alpha) = \frac{1}{2\pi\alpha^2} \cdot \exp\left(-\frac{d}{2\alpha^2}\right)$$

$$\alpha^* = \arg\max_{\alpha_{[n]}} \left\{ Variance\left(\sum_{i=1}^{K} F\left(D^N(Y,X_i)_{\alpha_{[n]}}\right)\right)\right\}$$

$$\alpha_{[n+1]} = \alpha_{[n]} + \varepsilon$$

$$\alpha_{[n=0]} = 0.01$$

$$\varepsilon = 0.01$$

The LVM Algorithm presents two main advantages: it does not require external parameters and it is based on Bayesian theory.

4.1.10   The MexProb Algorithm

The MexProb algorithm (Buscema et al. 2009a) was created to manage all the parameters usually employed in location theory algorithms within only one equation:

$\varphi =$ the connection strength among the points;
$B =$ the diameter of the protection zone, when such a zone is required;
$d =$ the distances among points, in any metric;
α and $\alpha^* =$ the width (and the optimal width) of the bell of the decay function.

The MexProb algorithm, moreover, calibrates all these parameters by itself, by maximizing the variance of its scalar field, iteratively:

$$F\big(D^N(Y,X_i)\big) = F\big(D^N(Y,X_i)_{\alpha*}\big) = F(d_{\alpha*});$$

$$F(d_\alpha) = (\varphi - B) + \tfrac{d}{2\cdot\alpha^2} \cdot e^{-\frac{d}{2\cdot\alpha^2}}$$

$$\alpha^* = \underset{\alpha_{[n]}}{\arg\max}\left\{ Variance\left(\sum_{i=1}^{K} F\big(D^N(Y,X_i)_{\alpha_{[n]}}\big)\right)\right\};$$

$$\alpha_{[n+1]} = \alpha_{[n]} + \varepsilon;$$
$$\alpha_{[n=0]} = 0.01;$$
$$\varepsilon = 0.01.$$

4.2   The Topological Weighted Centroid Basic Concepts

We define the Topological Weighted Centroid (TWC) of a set of N entities in a bi-dimensional space as the center of mass of such entities, weighted by the relative proximity of each entity to the others; the exact meaning of proximity is defined below. Since the weight employed in the proximity measure is dependent on a parameter α, we introduce a specific curve, $TWC(\alpha)$, such that the TWC point corresponds to a specific value α^*.

The $TWC(\alpha)$ curve is defined by the following iterative process:

$$\alpha(0) = 0; \{\text{Starting value for } \alpha\} \tag{4.1}$$

$$d_{i,j} = \sqrt{\left(x_i - x_j\right)^2 + \left(y_i - y_j\right)^2}; \{\text{Euclidean distance between any couple of entities}\} \tag{4.2}$$

$$D = \max_{i,j} \{d_{i,j}\}; \{\text{Maximum distance among the assigned entities}\} \tag{4.3}$$

$$AC_x = \frac{1}{N}\sum_{i=1}^{N} x_i; AC_y = \frac{1}{N}\sum_{i=1}^{N} y_i; \{\text{Classical Arithmetic Centroid}\} \tag{4.4}$$

$$p_i(\alpha) = \frac{1}{N-1}\sum_{j=1, j\neq i}^{N} e^{-\frac{d_{i,j}}{D}\alpha};$$

$$\{\alpha \text{ - dependent relative proximity of each entity w.r.t. the others}\} \tag{4.5}$$

$$TWC_x(\alpha) = \frac{1}{\sum_{i=1}^{N} p_i(\alpha)}\sum_{i=1}^{N} p_i(\alpha) \cdot x_i;$$

$$\tag{4.6}$$

$$TWC_y(\alpha) = \frac{1}{\sum_{i=1}^{N} p_i(\alpha)}\sum_{i=1}^{N} p_i(\alpha) \cdot y_i; \{\alpha \text{ - dependent TWC}\}$$

$$\alpha(t+1) = \alpha(t) + \Delta\alpha; \{\text{Increment of } \alpha\} \tag{4.7}$$

The TWC point is found on the curve by means of the following equations:

$$d_{AC,TWC}(\alpha) = \sqrt{(AC_x - TWC_x(\alpha))^2 + (AC_y - TWC_y(\alpha))^2};$$

$$\{\text{Distance between AC and TWC}(\alpha)\} \tag{4.8}$$

$$d_{AC,TWC}(\alpha^*) = \text{FL\_Max}_{\alpha} \{d_{AC,TWC}(\alpha)\};$$

$$\left\{\text{First Local Maximum distance between AC and TWC}(\alpha),\right.$$

$$\left.\text{obtained for } \alpha = \alpha^*\right\}$$

$$\tag{4.9}$$

$$p_i = \frac{1}{N-1} \sum_{j=1, j \neq i}^{N} e^{-\frac{d_{i,j}}{D}\alpha^*}; \{\text{Proximity of each entity to the others}\} \qquad (4.10)$$

$$TWC_x = \frac{1}{\sum_{i=1}^{N} p_i} \sum_{i=1}^{N} p_i \cdot x_i;$$

$$(4.11)$$

$$TWC_y = \frac{1}{\sum_{i=1}^{N} p_i} \sum_{i=1}^{N} p_i \cdot y_i; \{\text{Topological Weighted Centroid}\}$$

It can be easily seen that, as the iterative process unfolds, the following characteristic pattern of evolution emerges:

$$TWC(0) = AC \qquad (4.12)$$

$$\lim_{t \to \infty} p_i(t) = 0 \qquad (4.13)$$

$$TWC_k^0 = \lim_{t \to \infty} TWC_k(t) = \frac{k_{i_m} + k_{j_m}}{2}; d_{i_m, j_m}$$

$$= \min_{i,j} \{d_{i,j}\}, \text{ supposing unique min distance } ; k = x, y \qquad (4.14)$$

$$TWC_k^0 = \lim_{t \to \infty} TWC_k(t) = \sum_{r=1}^{R} \frac{k_{i_r} + k_{j_r}}{2 \cdot R}; d_{i_m, j_m} = d_{i_r, j_r} = \min_{i,j} \{d_{i,j}\},$$

$$r = 1, \ldots, R; \text{ supposing R min equal distances } ; k = x, y \qquad (4.15)$$

Equation 4.12 indicates that, at the beginning of the iterations, $TWC(\alpha)$ coincides with the classical arithmetic centroid; (4.14) and (4.15) show that, during the process, $TWC(\alpha)$ moves from the arithmetic centroid towards the center of mass of the points with minimum reciprocal distance. In case there is a single couple, (4.14) applies, whereas (4.15) applies for the generic case of R couples with equal minimum reciprocal distance.

From the previous considerations it is obvious that, for every set of points with given coordinates, it is possible to calculate the TWC. According to (4.1, 4.2, 4.3, 4.4, 4.5, 4.6, 4.7, 4.8, 4.9, and 4.10), the TWC appears to be a modification of the classical Arithmetic Centroid (AC), based on the concept of proximity of a point to the others. Equation 4.5 determines, for each assigned point, its proximity in terms of positioning in a more or less dense region of points. As observed, for each different value assumed by the α parameter during the iteration, we have a shifting position for the $TWC(\alpha)$. We chose to define the TWC point as the position assumed by $TWC(\alpha)$ for $\alpha = \alpha^*$, when the distance from the AC reaches its first local maximum.

The α parameter has a crucial role in this process, and is not simply an instrumental parameter. When $\alpha = 0$, each point acts independently, i.e. it does not 'feel' the presence of the others, and in this case the TWC and the AC collapse into each other. When α becomes positive (and growing), the area of influence of each point in the distribution widens, so that the proximity of the other points begins to matter, and as a consequence the position of $TWC(\alpha)$ changes as the degree of proximity of each point is weighted by α. The process stops when further enlargements of the area of influence of each point no longer affect the position of $TWC(\alpha)$. When this happens, the position of the moving $TWC(\alpha)$ is stuck at TWC^0, i.e., in the point where the relative proximity between the closest points has been α-calibrated in order to dominate the entire space.

Thus, by increasing α we draw out a correspondent path for $TWC(\alpha)$. This path tracks the position of the centroid weighing the proximity of the points while their influence area expands. It is evident that the more interesting positions for $TWC(\alpha)$ are neither the origin nor the end of the path. In particular, the first local maximum distance position seems to possess interesting properties.

When every entity has the same probability of occupying a place near the others, the space has no semantics and the TWC and the AC are coincident. This can happen in two limit cases: When the distances among the entities is always the same, and when the summation of the distances from one entity to the others is the same for every entity. These are two cases of perfect symmetry.

4.2.1   The Self Topological Weighted Centroid

If, at this point, we make a little change to (4.5) including the distance that each entity has from itself, the dynamics of the whole process change dramatically. To emphasize this apparently minor but crucial modification, in the following equations we prefer to re-label the name of the α parameter to β. Consistently, we do not speak anymore of TWC, but we introduce a new quantity called Self Topological Weighted Centroid ($STWC$). In this case, the weight used to calculate proximity now depends on the β parameter and, accordingly, a new $STWC(\beta)$ curve is considered, with a specific value β^* that fixes the STWC point, and that is determined following the very same logic as for TWC(α).

The algorithm to determine the $STWC(\beta)$ can be defined as an iterative process as follows:

$$\beta(0) = 0; \{\text{Starting value for } \beta\} \qquad (4.16)$$

$$d_{i,j} = \sqrt{\left(x_i - x_j\right)^2 + \left(y_i - y_j\right)^2}; \{\text{Euclidean distance between any couple of entities}\}$$

$$(4.17)$$

$$D = \max_{i,j} \{d_{i,j}\}; \{\text{Maximum distance among the assigned entities}\} \quad (4.18)$$

$$AC_x = \frac{1}{N} \sum_{i=1}^{N} x_i; \; AC_y = \frac{1}{N} \sum_{i=1}^{N} y_i; \{\text{Classical Arithmetic Centroid}\} \quad (4.19)$$

$$p_i(\beta) = \frac{1}{N} \sum_{j=1}^{N} e^{-\frac{d_{i,j}}{D}\beta};$$

$\{\beta\text{-dependent relative proximity of each entity w.r.t. the others, itself included}\}$

$$(4.20)$$

$$STWC_x(\beta) = \frac{1}{\sum\limits_{i=1}^{N} p_i(\beta)} \sum_{i=1}^{N} p_i(\beta) \cdot x_i;$$

$$STWC_y(\beta) = \frac{1}{\sum\limits_{i=1}^{N} p_i(\beta)} \sum_{i=1}^{N} p_i(\beta) \cdot y_i; \{\beta\text{ - dependent STWC}\} \quad (4.21)$$

$$\beta(t+1) = \beta(t) + \Delta\beta; \{\text{Increment of } \beta\} \quad (4.22)$$

The actual STWC is then determined by means of the following equations:

$$d_{AC,STWC}(\beta) = \sqrt{(AC_x - STWC_x(\beta))^2 + (AC_y - STWC_y(\beta))^2};$$
$$\{\text{Distance between AC and STWC}(\beta)\} \quad (4.23)$$

$$d_{AC,STWC}(\beta^*) = \max_{\beta} \{d_{AC,STWC}(\beta)\};$$

$\{\text{Maximum distance between AC and STWC}(\beta), \text{ obtained for } \beta = \beta^*\} \quad (4.24)$

$$p_i = \frac{1}{N} \sum_{j=1}^{N} e^{-\frac{d_{i,j}}{D}\beta^*}; \{\text{Proximity of each entity to the others, itself inclued}\}$$

$$(4.25)$$

$$STWC_x = \frac{1}{\sum\limits_{i=1}^{N} p_i} \sum_{i=1}^{N} p_i \cdot x_i; \; STWC_y = \frac{1}{\sum\limits_{i=1}^{N} p_i} \sum_{i=1}^{N} p_i \cdot y_i; \quad (4.26)$$

$\{\text{Self Topological Weighted Centroid}\}$

The corresponding pattern of evolution is characterized as follows:

$$\lim_{t \to \infty} p_i(t) = 1/N \tag{4.27}$$

$$\lim_{t \to \infty} STWC_k(t) = STWC_k(0) = AC_k \tag{4.28}$$

These equations show the convergence properties of the algorithm; in particular, one can notice that the $STWC(\beta)$ dynamics describe a loop, starting from, and ending with, the classical arithmetic centroid.

From (4.24), we see that β^* is the parameter value at which the absolute maximum is reached. So, β^* is a critical point: For any $\beta > 0$ up to β^*, the distance between $STWC(\beta)$ and AC typically increases as each $p_i(\beta)$ (see (4.20)) is greater than 0, smaller than 1 and, excluding the non-generic case of a null semantics, different from the others. Further increases of β beyond β^* cause the distance of each point from itself to dominate on the other distances and consequently $\partial STWC(\beta)$ and AC slowly approach each other.

With respect to the proximity concept defined for the TWC as in (4.5), in the one defined here as in (4.20), the proximity to other points is valued only as an additive component to the presence of the point itself. As β moves along its range, the corresponding changes in $STWC(\beta)$ position the curve such as to weigh in terms of β the area of influence corresponding to this kind of proximity, thereby mediating the presence of the other points with that of the reference point itself. At the β^* value, such mediation effect causes the extension of the area of influence to diverge as much as possible from the ones of the areas that correspond to the (coincident) initial and final positions along the $STWC(\beta)$ curve.

4.2.2 The Proximity Scalar Field

Let us now discretize our bi-dimensional plane by assigning to it a (discrete) set of geometrical points, e.g., a grid. Each geometrical point may, or may not, be covered by one of the entities that express the phenomenon under study. On the basis of the proximity values p_i described above, we can define a scalar field in the discrete plane, and compute a specific value for each of its points. We call it the *Proximity Scalar Field*. The value assigned to each geometrical point of the discretized plane represents the proximity degree of that point to the observed entities. A useful way to construct such a scalar field is that, whenever the coordinates of a geometrical point happen to coincide with those of an entity, the scalar field tends to assume a relatively high value; moreover, whenever there is also a significant concentration of other entities close to the 'matching' entity/point, the scalar field value increases further. To sum up, we define our scalar field onto a discrete covering of the plane in such a way that it constitutes a sort of density measure for entities.

Comparing the two notions of proximity as defined in (4.5) and (4.20), it is easy to conclude that the reflexive one, i.e., the one based on the β^* value, is more appropriate in this case, in that reflexive distance is a substantial component of our definition of a proximity scalar field.

We therefore define the scalar field for the assigned entities in terms of the following equations:

$$N; \{\text{Number of entities generating the field}\} \tag{4.29}$$

$$M; \{\text{Number of points of the discretized space}\} \tag{4.30}$$

$$i, j \in \{1, 2, \ldots, N\}; \{\text{Entities indexes}\} \tag{4.31}$$

$$k \in \{1, 2, \ldots, M\}; \{\text{Space points index}\} \tag{4.32}$$

$$d_{i,j} = \sqrt{\left(x_i - x_j\right)^2 + \left(y_i - y_j\right)^2}; \tag{4.33}$$
$$\{\text{Euclidean distance between any couple of entities}\}$$

$$D = \max_{i,j} \{d_{i,j}\}; \{\text{Maximum distance among the assigned entities}\} \tag{4.34}$$

$$m_{k,j} = \sqrt{\left(x_k - x_j\right)^2 + \left(y_k - y_j\right)^2};$$
$$\{\text{Euclidean distance between any couple of space points and entities}\}$$
$$\tag{4.35}$$

$$p_k = \frac{1}{N} \sum_{j=1}^{N} e^{-\frac{m_{k,j}}{D}\beta^*};$$
$$\{\text{Proximity of a point to the entities, with } \beta^* \text{ as maximizing value of } d_{\text{AC,STWC}}\}$$
$$\tag{4.36}$$

From the above equations, it is evident that our *Proximity Scalar Field* is a differentiable function, and at any point it yields a value that represents the point's proximity to the given entities, including the one possibly coinciding with the point itself. As for the *STWC*, proximity is weighed using an area of influence that maximizes the distance of the weighed centroid from the arithmetic one. In other words, we define a concept of centrality that re-maps the space in such a way to consistently maximize the semantic distance with respect to the 'context-neutral' (i.e., arithmetic) idea of centrality.

4.2.3   The Gradient of the Scalar Field

Once a scalar field is generated from the geometrical points, it may be useful to calculate the gradient of this field, to transform again the scalar field into a conservative vector field. In this way, we can assign to each entity or to a new point a virtual path, that defines a dynamics on the whole plane, that is implicit in the semantics of the generated target space.

In order to compute the x and y components of the gradient of this scalar field, we have applied the Sobel operator (Duda and Hart 1973; Jaehne et al. 1999) to assign a measure of relevance to the points around the reference one. The masks to compute the gradient in the form of the Sobel operator are:

−1	0	1
−2	0	2
−1	0	1

for the x component of the gradient, G^*_x, and

1	2	1
0	0	0
−1	−2	1

for the y component of the gradient, G^*_y. Therefore, the equations computing these components are:

$$G^*_x = \left(p_{x+1,y+1} + 2 \cdot p_{x+1,y} + p_{x+1,y-1}\right) - \left(p_{x-1,y+1} + 2 \cdot p_{x-1,y} + p_{x-1,y-1}\right)$$

$$\text{(4.37)}$$

$$G^*_y = \left(p_{x+1,y+1} + 2 \cdot p_{x,y+1} + p_{x-1,y+1}\right) - \left(p_{x+1,y-1} + 2 \cdot p_{x,y-1} + p_{x-1,y-1}\right)$$

The gradient intensity and its local direction represent the *variation of the proximity* with respect to the entities situated in each point of the plane. It can be seen as the force of the *target space* at each point, that would act on an entity placed at that point, due to the potential information content generated by the anisotropy of the space in terms of proximity – that is to say, the different degree of proximity that each entity has with respect to each other, and with respect to any geometrical point of the *target space*. In such a way, the *semantics* of a target space could be regarded as the potential energy of information of that space, that could act upon the state of any newly added entity according to its inertia, which could be seen in turn as the own information content of that entity.

4.2.4   Relative Topological Weighted Centroid and Paths from AC to Entities

We can define a centroid relative to any specific i-th entity. Actually, to weigh the entity coordinates we can consider only the distances among a given entity and all the others, itself included, but excluding all the distances among other entities. The Relative Topological Weighted Centroid for the i-th entity will be indicated by TWC_i. Also for this quantity, a corresponding curve $TWC_i(\gamma_i)$ is defined, according to the by now familiar logic, and dependent on the parameter γ_i, such that a specific value γ_i^* identifies the TWC_i.

The $TWC_i(\gamma_i)$ curve is algorithmically defined as follows:

$$\gamma_i(0) = 0; \{\text{Starting value for } \gamma_i\} \tag{4.38}$$

$$d_{i,j} = \sqrt{\left(x_i - x_j\right)^2 + \left(y_i - y_j\right)^2};$$
$$\{\text{Euclidean distance between the i-th and the j-th entity}\} \tag{4.39}$$

$$D = \max_{i,j}\{d_{i,j}\}; \{\text{Maximum distance among the assigned entities}\} \tag{4.40}$$

$$AC_x = \sum_{i=1}^{N} x_i; AC_y = \sum_{i=1}^{N} y_i; \{\text{Classical Arithmetic Centroid}\} \tag{4.41}$$

$$p_{i,j} = e^{-\frac{d_{i,j}}{D}\gamma_i}; \{\gamma \text{ - dependent relative proximity between the i-th and the j-th entities}\} \tag{4.42}$$

$$TWC_{i_x}(\gamma_i) = \frac{1}{\sum_{j=1}^{N} p_{i,j}(\gamma_i)} \sum_{j=1}^{N} p_{i,j}(\gamma_i) \cdot x_j;$$

$$TWC_{i_y}(\gamma_i) = \frac{1}{\sum_{j=1}^{N} p_{i,j}(\gamma_i)} \sum_{j=1}^{N} p_{i,j} \cdot y_j(\gamma_i); \{\gamma \text{ - dependent } TWC_i\} \tag{4.43}$$

$$\gamma_i(t+1) = \gamma_i(t) + \Delta\gamma_i; \{\text{Increment of } \gamma_i\} \tag{4.44}$$

The TWC_i quantity is then identified by the equations below:

$$d_{AC,TWC_i}(\gamma_i) = \sqrt{(AC_x - TWC_{i_x}(\gamma_i))^2 + (AC_y - TWC_{i_y}(\gamma_i))^2};$$
$$\{\text{Distance between AC and TWC}_i(\gamma_i)\} \tag{4.45}$$

$$d_{AC,TWC_i}(\gamma_i^*) = \max_{\gamma_i} \{d_{AC,TWC_i}(\gamma_i)\};$$

$\{$Maximum distance between AC and $TWC_i(\gamma_i)$, obtained for $\gamma_i = \gamma_i^*\}$

(4.46)

$$p_{i,j} = e^{-\frac{d_{i,j,\gamma_i^*}}{D}};$$

$\{$Proximity between the i-th and the j-th entities, with γ_i^*

as maximizing value of $d_{AC,TWC_i}\}$ (4.47)

$$TWC_{i_x} = \frac{1}{\sum_{j=1}^{N} p_{i,j}} \sum_{j=1}^{N} p_{i,j} \cdot x_j;$$

(4.48)

$$TWC_{i_y} = \frac{1}{\sum_{j=1}^{N} p_{i,j}} \sum_{j=1}^{N} p_{i,j} \cdot y_j; \{\text{Relative Topological Weighted Centroid}\}$$

and the corresponding dynamics is characterized as:

$$TWC_i(0) = AC \qquad (4.49)$$

$$\lim_{t \to \infty} p_{i,j}(t) = 0 \qquad (4.50)$$

$$\lim_{t \to \infty} TWC_{i_k}(t) = k_i; k = x, y \qquad (4.51)$$

As stated by (4.38), at the beginning of the iterations $TWC_i(\gamma_i)$ coincides with the AC. As the parameter γ_i increases, the centroid moves towards the i-th entity, eventually reaching it, as shown by (4.51).

For the $TWC_i(\gamma_i)$ what is interesting is obviously not the endpoint of the path, but the path itself, from the center of mass (AC) to the entity, as the γ_i parameter increases. Such trajectory will be a straight line when the entities have a distribution with null semantics. It will assume, instead, a curved shape when the proximity field has a specific warped geometry. This deformation shows how the path from the AC to the given entity is modified by the attraction effects caused by the proximity of all the other entities in the space.

4.2.4.1 Paths Between Entities

It is relatively easy to translate the path from the AC to the entities into direct paths connecting entities among each other. Their parametric coordinates are given as follows:

$$C_{(i,j)_x}(s) = x_j + \left(TWC_{i_x}\left(\gamma_i^* \cdot s\right) - TWC_{j_x}\left(\gamma_j^* \cdot s\right)\right);$$

$$C_{(i,j)_y}(s) = y_j + \left(TWC_{i_y}\left(\gamma_i^* \cdot s\right) - TWC_{j_y}\left(\gamma_j^* \cdot s\right)\right); 0 \le s \le 1 \qquad (4.52)$$

In this way, we can draw a complete regular undirected graph, whose vertices are the entities and whose arcs are in general curvilinear, moving from the j-th entity (for $s = 0$) to the ith (for $s = 1$). Similarly to the paths from the AC to the entities, the trajectory will be a straight line in case of a null semantics, but will assume a characteristic shape when the proximity to the other entities varies along the space crossed by the path joining the two entities.

The **Non Linear Minimum Spanning Tree** (NL-MST) of this graph could be a good representation of the trajectories connecting all the entities.

4.2.5   The Scalar Field of the Trajectories

We are now in the position to be able to calculate the degree of proximity of each point of the plane to every point of the above defined trajectories:

$$N; \{\text{Number of points composing all the trajectories, in the discrete space}\}$$
$$(4.53)$$

$$M; \{\text{Number of points of the discrete space}\} \qquad (4.54)$$

$$i, j \in \{1, 2, \dots, N\}; \{\text{Trajectories points indexes}\} \qquad (4.55)$$

$$k \in \{1, 2, \dots, M\}; \{\text{Space points index}\} \qquad (4.56)$$

$$d_{i,j} = \sqrt{\left(x_i - x_j\right)^2 + \left(y_i - y_j\right)^2}; \{\text{Euclidean distance between any couple}$$

$$\text{of trajectories points}\}$$
$$(4.57)$$

$$D = \max_{i,j} \{d_{i,j}\}; \{\text{Maximum distance among the trajectories points}\} \qquad (4.58)$$

$$m_{k,j} = \sqrt{\left(x_k - x_j\right)^2 + \left(y_k - y_j\right)^2}; \{\text{Euclidean distance between any couple}$$

$$\text{of space points and trajectories points}\}$$
$$(4.59)$$

$$p_k = \frac{1}{N} \sum_{j=1}^{N} e^{-\frac{m_{k,j}}{D}\beta^*}; \{\text{Proximity of a point to the trajectory points,} \tag{4.60}$$

with β^* as maximizing value of $d_{AC,STWC}\}$

This field yields, in any point, a value that represents the proximity of the point itself to the trajectories connecting entities. As for the scalar field, proximity is weighed by the β^* parameter, as defined for the *STWC*.

Part B – Syntax of Physical Space: Applications

4.3 Introduction

In the first part of the paper, we have briefly reviewed the essential tools that we need to carry out a *relational* analysis of complex spatial dynamics, based on the idea that the dynamics is essentially governed by a semantics of space that arises from the negotiation among the physical entities that constitute the observable correlate of the underlying phenomenon. Most of the toolbox (and of the respective concepts) that we have introduced is likely to sound relatively new to the reader who would probably call for a much more extensive discussion and for further argument and clarification. Likewise, the same can be said about the general philosophy and the specific meaning of the various parts of the TWC methodology. However, rather than taking this route, in order to maintain the length of the paper within reasonable limits, we have opted for a different solution: That of illustrating systematically the application of the methodology by means of examples motivated by a few issues relating to different disciplines. Most of these examples are not trivial problems for they contain several original results on disciplinarily relevant issues, and are presented here for the first time. We hope that, by going through the mechanics of the analytical procedures in the various examples below, the reader will be able to return to the theoretical part with more insight and better prepared to understand the theoretical complexities.

4.3.1 A Basic Example with a Very Weak Semantic

We introduce as an initial illustration a space with a very weak semantic. In fact, in this space, the summation of distances of each entity from the others is quite the same, and consequently the *AC* and the *TWC* are located approximately in the same position (Fig. 4.1): the software use for all simulations and graph come from Buscema (2007b, 2008).

Fig. 4.1 A simple artificial example

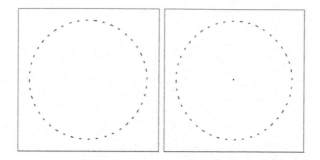

Fig. 4.2 The scalar field

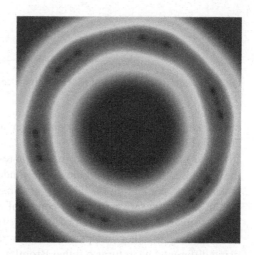

By means of 4.29, 4.30, 4.31, 4.32, 4.33, 4.34, 4.35, and 4.36, we can generate the proximity scalar field of the entities (Fig. 4.2):

The map shows that the degree of proximity to entities is quite evenly distributed around the circle. We can also notice the darker areas inside the red region, that correspond to the places where the original entities are denser. Using 4.38, 4.39, 4.40, 4.41, 4.42, 4.43, and 4.44 we can also generate the trajectories connecting each entity with the center of mass (*AC*) (Fig. 4.3). Also in this case, the connecting lines are mostly straight, because the proximity field around entities is quite the same for each.

4.3.2   Cocaine Trafficking in London, 2006

We now consider a completely different problem based on real world data and presenting a typical example of what is commonly meant as a 'hard' analytical task. We start from a map of the places in London where the Metropolitan Police arrested

Fig. 4.3 Trajectories connecting entities with the center of mass

people caught trafficking in cocaine over a four month period (Buscema et al. 2009b) (Fig. 4.4):

An examination of this map makes it relatively clear how the distribution of the entities is hardly believable to be random. In the following picture we can observe how, computing the TWC according to 4.1, 4.2, 4.3, 4.4, 4.5, 4.6, and 4.7, it can be seen to move from the center of mass (AC) down towards a group of entities situated in the borough of Bromley (Fig. 4.5).

According to other independent research (Buscema et al. 2009b), conducted using different data, it turns out that Bromley and the City of Westminster appear as two of the most representative boroughs in London as to cocaine trafficking. If we generate the scalar field of this distribution, using 4.29, 4.30, 4.31, 4.32, 4.33, 4.34, 4.35, and 4.36 we have an optimal representation of the proximity to cocaine trafficking in London: It is apparent that Bromley and the City of Westminster are the two foci of cocaine distribution (Fig. 4.6).

Now, projecting the same map according to the gradient, as in (4.37), we are also able to distinguish the different clusters of cocaine trafficking in London (Fig. 4.7):

At this point, it would be interesting to try and conjecture the most probable trajectories connecting all such drug dealing places (entities) together. We can try to answer to this question by building a curvilinear MST, based on the paths connecting entities, according to (4.52) (Fig. 4.8).

This graphical map could at first sight look too complex to be readable in any sensible way. To help the reader (and the analyst), we can provide an alternative representation obtained by employing the Maximally Regular Graph (MRG) algorithm, that adds to the MST representation all and only the links that are necessary to capture the fundamental connections between the observed entities (Buscema et al. 2008a; Buscema and Sacco 2008; Buscema and Grossi 2008; Buscema et al. 2008b; Buscema et al. 2009b, under review) (Fig. 4.9).

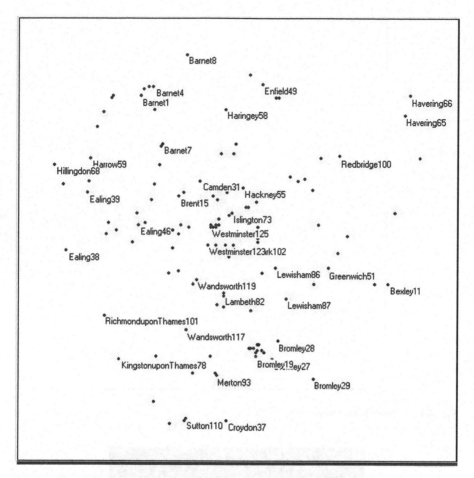

Fig. 4.4 Arrests for cocaine trafficking in London over a 4 month period

By looking at this graph, it is possible to infer that the cocaine delivery network in London starts from the borough of Bromley (TWC high in the north of the graph), has its wholesale hub in the borough of Merton and finally arrives, through Lambeth, at the big, final customer market of the City of Westminster and subsequently to the other boroughs of London (relatively minor final customer markets).

This example explains rather clearly what we mean by a relational spatial analysis, and to what extent our approach exploits a particular form of space semantics. The task of reconstructing the spatial articulation of cocaine trafficking in a large metropolitan area in London by looking only at events that locate the capture of drug dealers in a given time period looks somewhat hopeless at first sight. Nevertheless, it is clear that, insofar as a real drug trafficking network is operating across the London territory, it must have an inherent spatial organization that has to emerge to some degree from the observable data. The question becomes whether a

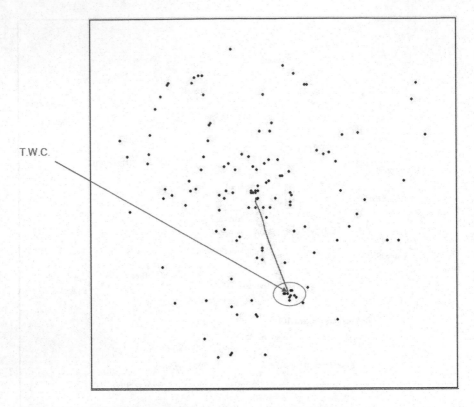

T.W.C.

Fig. 4.5 Centre of mass versus TWC

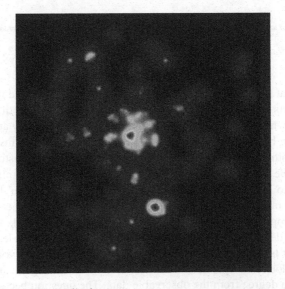

Fig. 4.6 Scalar field of the distribution

Fig. 4.7 Gradient map

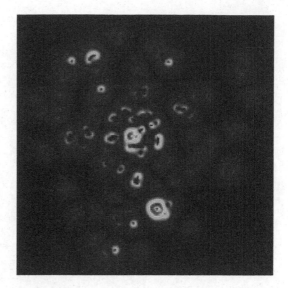

Fig. 4.8 Nonlinear MST

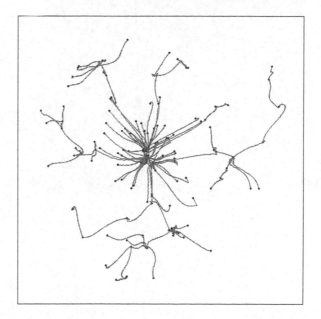

small scatter of points like the available database for this example may be enough to reconstruct the basic structural properties of the drug dealing network. If we consider the available data (the entities) as isolated points in space, i.e. as random draws from a given, underlying distribution, the task is really hopeless. But if, on the contrary, such data are thought as context-sensitive, i.e., a given event (the arrest of a drug dealer) produces complex *spatial* effect on the entire geography and organization of

M. Buscema et al.

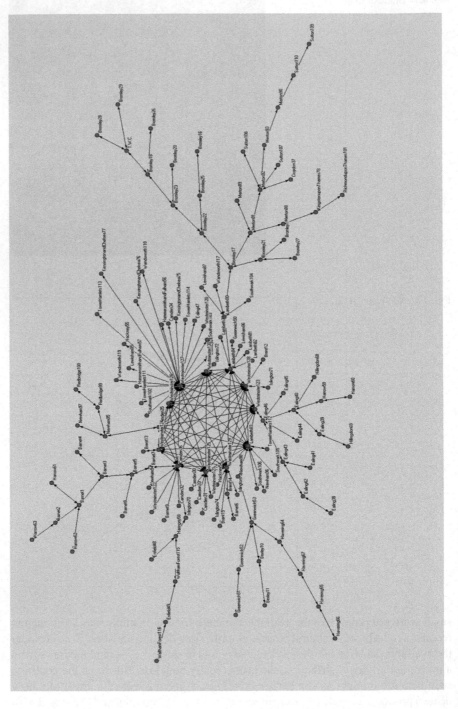

Fig. 4.9 Maximally regular graph (MRG)

drug trafficking (like a propagation wave following a shock), then it is clear that the observed spatial distribution of entities carries meaning, i.e., is governed by an underlying *semantic* that can be reconstructed on the basis of a few well-chosen reference cases. In this perspective, it should also be clearer in what sense the various entities negotiate their position in the underlying, warped representational space; Some of the dealers are more tightly connected to some of the others, and thus the event of the capture of a given dealer produces asymmetrical effects in the whole network, and affects different, as yet uncaught dealers in different ways, thereby causing complex adjustments that will be reflected in the distribution of the next capture. This is the kind of hidden, structural information that we need to learn to exploit to be able to reconstruct the true semantics of the phenomenon under study.

4.3.3   The Russian Influenza in Sweden, 1889–1890

Using data from a study of the 1889–1890 Russian flu in Sweden, this example shows how the application of the TWC methodology may help to identify the source of an epidemic outbreak in surveillance analyses (for a complementary study on the sources of outbreaks, see Buscema et al. (2009a)).

In 1890, immediately after the outbreak of Russian influenza, all Swedish doctors were asked to provide information about the start and the peak of the epidemic, and the total number of cases in their region, as well as to fill in a questionnaire on the number, sex and age of infected persons in the households they visited. General answers on the epidemic were received from 398 physicians, and data on individual patients were available for more than 32,600 persons. From such answers, a table was compiled and a map was drawn in 1890, indicating when the influenza first appeared at different locations. To support the contagiousness theory, an analysis of the connections between the spatial dynamics of the disease from the onset of its outbreak and the structure of the railway network was carried out. In the first week of December 1889, in turned out that 12 of the 13 affected places outside Stockholm had railway stations. Linroth (1890) demonstrated that, by December 20 of the same year, 82% of reporting places with a railway station and 47% without one had been affected. The dissemination was very fast and the local epidemics developed at a pace that in some cases was described as explosive. Due to the general susceptibility, the short incubation time and the difficulty to detect the very first cases, more evidence was needed to scientifically ascertain that the influenza was indeed contagious. Linroth was however of the opinion that the many individual testimonies describing how the infection was transferred directly from infected persons justified the hypothesis: Influenza is a contagious disease.

In a recent GIS study (Skog et al. 2008), Linroth's original tables were converted into an Excel format and to dot maps. The figure below, taken from (Skog et al. 2008), illustrates the progression of the epidemic across time. Dots represent the places where at least one case of infection had been reported to date, starting from the last week of November 1889. The railroad network is also shown (Fig. 4.10).

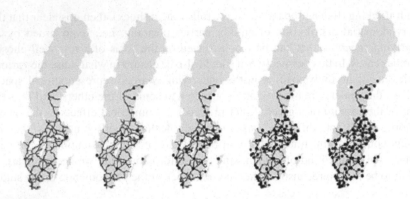

Fig. 4.10 Influenza epidemic spread versus railways system in Sweden, 1889

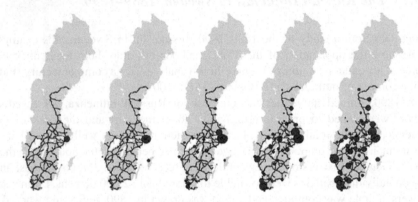

Fig. 4.11 Four-weeks progression of the influenza

In this other figure, dots sizes are proportional to the number of reported cases. Each map represents one week, starting from the last week of 1889 (first left) (Fig. 4.11).

The figure below shows a magnification of dot maps at week zero and week three of the Swedish epidemic spread (Fig. 4.12).

We have worked on the scatter of dots of week 3, using the coordinates of the points corresponding to the 44 locations interested by the outbreak in the third week, in order to calculate the appropriate mathematical quantities described in the first part of the article. The coordinates were derived from the dot maps. Figure 4.13 shows the 44 dot maps obtained from the original coordinates, from which we have implemented the TWC methodology.

The source of the epidemic corresponds to a location between points 25 and 30 in our dot maps.

As it is shown in Fig. 4.14, starting from the Euclidean centroid, the system positions the TWC α^* exactly between points 25 and 30, the source of the epidemic spread.

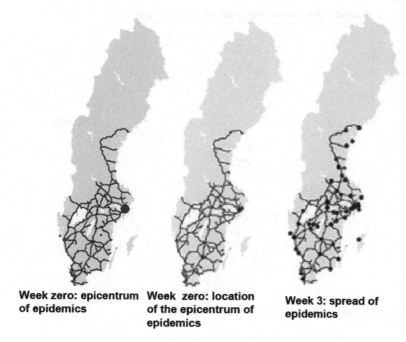

Week zero: epicentrum
of epidemics

Week zero: location
of the epicentrum of
epidemics

Week 3: spread of
epidemics

Fig. 4.12 Week zero versus week three spread

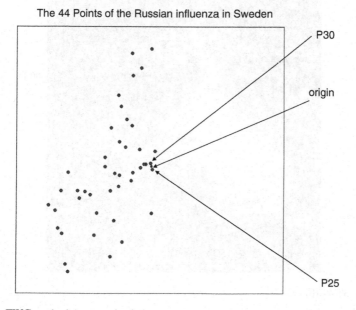

The 44 Points of the Russian influenza in Sweden

P30

origin

P25

Fig. 4.13 TWC methodology on the dot map

The TWC (Alfa) moves close the origin of the epidemic.

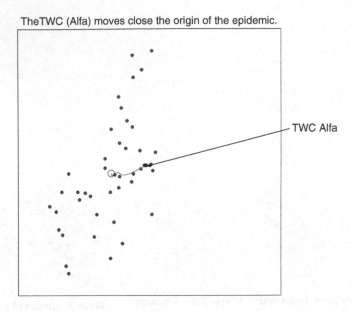

Fig. 4.14 TWC versus the actual epidemic source point

The TWC (Beta) map points the place of the origin of the epidemic and the probability of its diffusion

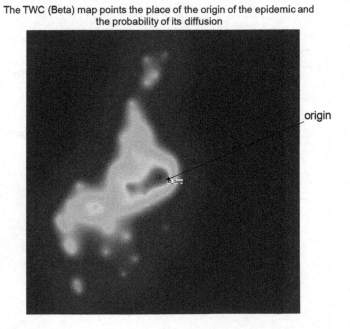

Fig. 4.15 STWC(β) map

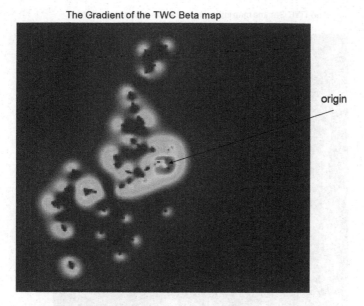

Fig. 4.16   Gradient of the STWC(β) map

In Fig. 4.15, we report the STWC(β) map, again correctly identifying the point of origin of the epidemic and additionally determining the probability of its diffusion.

The gradient of the STWC (β) map yields the likely speed of diffusion of the epidemic from its origin (Fig. 4.16). It is noteworthy that our emergent dot map does anticipate quite closely the real spread observed in weeks 4 and 5.

The TWC (γ) map picks even more precisely the origin of the epidemic and the actual probability of its diffusion (Fig. 4.17).

The gradient of the TWC(γ) map provides again the likely speed of diffusion of the epidemic from its origin (Fig. 4.18).

Finally, with TWC(γ) the possible paths of diffusion of the epidemic are drawn out (Fig. 4.19a). It is interesting to notice that such paths are very close to the actual railway network as reported in the maps in Fig. 4.19b.

In this other example, we find a different type of complex spatial dynamics: From capture to contagion. In fact, one might see the contagion dynamics in terms of the capture of the infected humans by the viral agent. But the logic of capture in the two cases is basically different, in that in the latter the number of predators is extremely high and capture occurs through physical proximity and contact, whereas in the former predators are few and capture occurs through careful intelligence and a dose of good luck. Nevertheless, we find that this entirely different kind of spatial phenomenon may also be addressed through the same battery of tools, and again providing a rather accurate reconstruction of the underlying structural forces at play. Again, this is a case where thinking of the phenomenon in 'relational' terms and trying to reconstruct its implicit semantics makes sense, and this time in a very intuitive way: Each case of contagion clearly determines an area of influence

The TWC Gamma map indicates more exactly the origin of the
epidemic and the probability of its diffusion

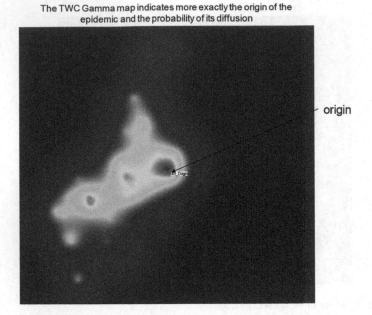

Fig. 4.17 TWC(γ) map

The Gradient of the TWC Gamma map indicates the possible speed
of diffusion of the epidemic from its origin

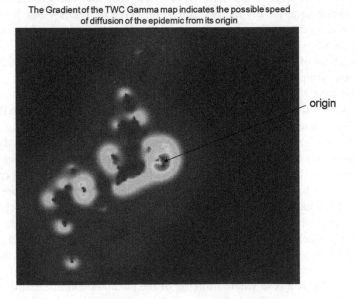

Fig. 4.18 Gradient of the TWC(γ) map

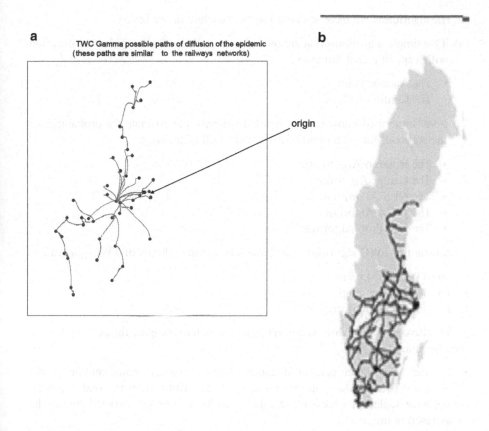

Fig. 4.19 (**a**) TWC(γ) paths; (**b**) Sweden railway networks

(morbility) that drives the appearance of further cases, the more or less so depending on a number of local circumstances (size of the family, type of housing, pre-existence of other diseases, etcetera). But also the long-range interactions determined by the railway system are mirrored in the results, and with an interesting level of accuracy: This is a clear illustration of the global character of the relational concept on which the TWC methodology is based.

4.3.4   Comparison with the Other Algorithms

The case of Russian influenza in Sweden is pretty well known. In this case, we know the exact location of the outbreak and of its dynamics, and thus we can carry out any kind of benchmark comparison test. Specifically, we can compare the performance of TWC algorithms with that of the other algorithms presented in Sect. 1.2.

The algorithms we have selected for this benchmark are seven:

(a) Two simple algorithms that are only able to work upon a minimization criterion with a specific cost function:

- The Anchor Point;
- The Centre of Mass;

(b) Five (some well-known, some new) algorithms that estimate the probability of the outbreak for each point of the convex hull of the map:

- The Rossmo Algorithm;
- The Canter Algorithm;
- The NES Algorithm;
- The LVM Algorithm;
- The MexProb Algorithm.

Among the TWC algorithms, we have selected three kinds of TWC quantities:

- The TWC(Alfa) point;
- The TWC(Beta) map;
- The TWC(Gamma) map.

We also considered various criteria against which to compare the performance of the algorithms:

- *Percent error*: The percent distance of the estimated point and/or of the estimated peak of the map generated by the algorithm from the real outbreak. We have scaled this measure in relation to the maximum distance among the assigned points.
- *Sensitivity*: The probability that each algorithm assigns on its characteristic map the point where the real outbreak occurred (only for the algorithms whose output is a probabilistic map).
- *Specificity*: If we divide the probability distribution into 9 equivalent bins from 0 to 1, we have 9 classes of probability; the percent of points of all the classes of the map where the real outbreak point is *not* located defines the specificity of the algorithm.

The table and the figures below show a general convergence of the more complex algorithms toward the real outbreak point:

Algorithms	Type of output	Distance from the outbreak – accuracy (%)	Sensitivity (%)	Specificity (%)
TWC(Gamma)	Probability Map (Peak)	1.67	90.13	99.89
TWC(Alfa)	Point	3.13		
Canter	Probability Map (Peak)	3.81	58.86	99.83
TWC(Beta)	Probability Map (Peak)	4.55	77.59	99.50

(continued)

Algorithms	Type of output	Distance from the outbreak – accuracy (%)	Sensitivity (%)	Specificity (%)
NES	Probability Map (Peak)	4.55	79.08	99.46
MexProb	Probability Map (Peak)	6.73	82.69	99.32
LVM	Probability Map (Peak)	6.77	84.57	98.14
Anchor Point:	Point	15.90		
Mean Center:	Point	17.44		
Rossmo	Probability Map (Peak)	19.42	80.18 %	96.36 %

Some considerations are in order.

- TWC(Alfa) is the best estimation among the algorithms with point-wise output, and it is the second best in the general ranking.
- TWC(Gamma) peak points out the real outbreak with a very consistent accuracy, high sensitivity and very high specificity.
- Canter algorithm is quite accurate but its sensitivity is low. Rossmo algorithm is not accurate and its specificity is lower than the others.
- The other algorithms are positioned in middle range of performance: A reasonable accuracy, an acceptable sensitivity and a high specificity.

In the following Figs. 4.20, 4.21, 4.22, 4.23, 4.24, and 4.25 we visualize our results, sticking to the following conventions:

- The colours from dark blue to dark red indicate the estimation of the operation point location from 0.0 to 1.0;
- The name of the algorithm is at the top of the figure;
- The white point represents the peak of the map;
- The label "45_Origin" represents the real outbreak.

Part C – Semantics of Physical Space: Theory

4.4 The Auto Contractive Map and Qualitative Information

We can synthetically refer to the whole set of the equations and the quantities presented thus far as the *TWC Methodology*. The TWC Methodology works upon a dataset of N two-dimensional (or alternatively, three-dimensional) entities:

$$DATASET : \{x_n, y_n\}_{n=1}^{N} \text{ where } x_n \text{ and } y_n \text{ are the (geographical) coordinates}$$
$$\text{of each entity within the 2 - dimensional space.}$$

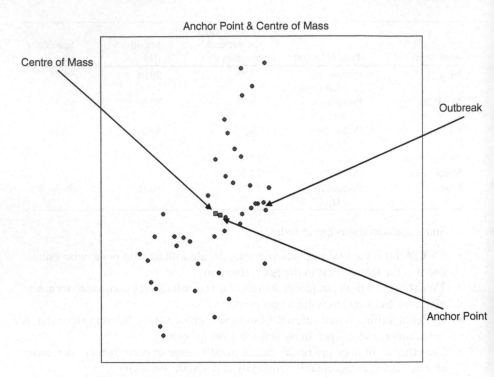

Fig. 4.20 Anchor point and centre of mass

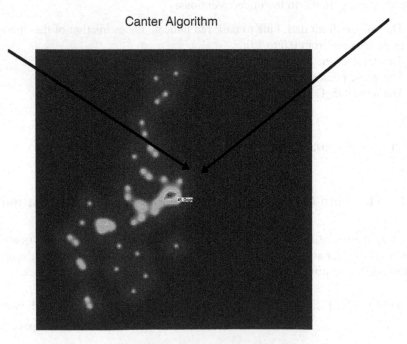

Fig. 4.21 Topological weighted centroid

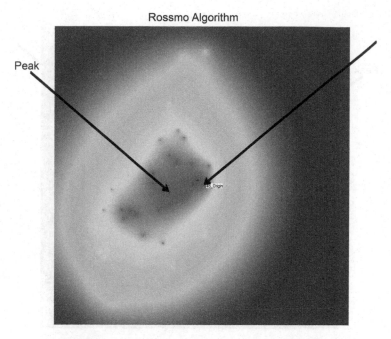

Fig. 4.22 Rossmo algorithm

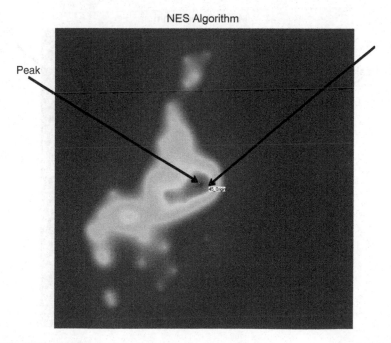

Fig. 4.23 NES algorithm

LVM Algorithm

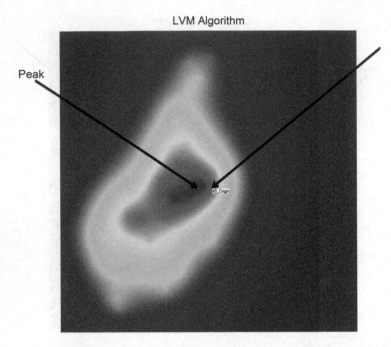

Fig. 4.24 LVM algorithm

MexProb Algorithm

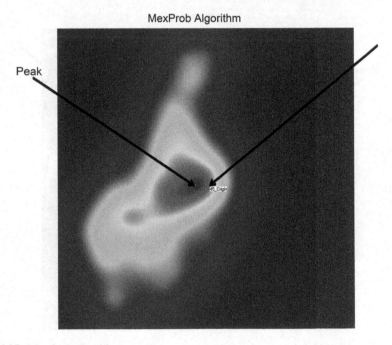

Fig. 4.25 Mex Prob algorithm

We can increase the complexity of the dataset, however, by attaching to each entity a vector of characteristics:

$$NEW\_DATASET : \{x_n, y_n, \{c_{n,p}\}_{p=1}^{P}\}_{n=1}^{N}$$

where c_n, p are the abstract characteristics of each entity in a suitable M-dimensional space.

With this additional set of non-spatial characteristics, we are passing from a syntactic to a fully semantic level of analysis, that is to say, to the semantics of space. If we want to merge all this information into a common representational space, we have to compute how the extra characteristics bring about modifications in the classic Euclidean space matrix, given the (Euclidean) coordinates x_n and y_n. To this purpose, we will process all the $c_{n,p}$ characteristics pertaining to each entity by means of an Auto-Contractive Map (Auto-CM). This ANN is a special Auto-Associative algorithm that is able to estimate the non linear association among the entities, given their respective sets of characteristics (Buscema et al. 2008a; Buscema and Sacco 2008; Buscema et al. 2008b; Buscema 2007a, c). Its basic equations are as follows:

Legend :

$C =$ constant (typically $C = D_{Max}^{[Eu]}$);

$D_{Max}^{[Eu]} = \max\limits_{i,j} \{d_{i,j}\}$;

$d_{i,j} =$ classic Euclidean distance between entities using the x and y coordinates;

$p \in P$ the p - th characteristic;

$i, j \in N$ indexes for the entities; $[z]$ = index for any cycle of the Auto - CM;

$In_{p,i}^{[z]} =$ the p - th feature of the i - th entity at cycle z;

$In_{p,i}^{[z]} \in [0, 1]$;

$v_i^{[z]} =$ weight of any entity at any cycle;

$w_{i,j}^{[z]} =$ weight between any couple of entities at any cycle;

$Hid_j^{[z]} =$ Hidden activation of each entity at any cycle;

$Net_i^{[z]} =$ Net input to any entity at any cycle;

$Out_i^{[z]} =$ Output of any entity at any cycle.

$$Hid_i^{[z]} = In_{p,i}^{[z]} \cdot \left(1 - \frac{v_i^{[z]}}{C}\right); \qquad\qquad (4.61)$$

$$v_i^{[z+1]} = v_i^{[z]} + \left(In_{p,i}^{[z]} - Hid_i^{[z]}\right) \cdot \left(1 - \frac{v_i^{[z]}}{C}\right) \cdot In_{p,i}^{[z]}; \qquad (4.62)$$

$$Net_i^{[z]} = \sum_j^N Hid_j^{[z]} \cdot \left(1 - \frac{w_{i,j}^{[z]}}{C}\right); \qquad (4.63)$$

$$Out_i^{[z]} = Hid_i^{[z]} \cdot \left(1 - \frac{Net_i^{[z]}}{C}\right); \qquad (4.64)$$

$$w_{i,j}^{[z+1]} = w_{i,j}^{[z]} + \left(Hid_i^{[z]} - Out_i^{[z]}\right) \cdot \left(1 - \frac{w_{i,j}^{[z]}}{C}\right) \cdot Hid_j^{[z]}; \qquad (4.65)$$

The Auto-CM convergence criterion is fixed as:

$$\sum_i^N \Delta v_i = 0; \text{ where } \Delta v_i = \left(In_{p,i}^{[z]} - Hid_i^{[z]}\right) \cdot \left(1 - \frac{v_i^{[z]}}{C}\right) \cdot In_{p,i}^{[z]} \text{ and } \Delta v_i \geq 0. \quad (4.66)$$

(for the demonstration of these conditions see Buscema [5,6,9]).

The trained **w** weights matrix encodes all of the nonlinear relationships among the N entities. With a simple operation, we transform the **w** matrix into a *semi-distance*:

$$d_{i,j}^{[AutoCM]} = C - w_{i,j} \; i \neq j. \qquad (4.67)$$

The weights matrix **w** *also* defines the local fan-out strength of each entity. This information is important to establish the influence of each entity in this new metric:

S_i = Local strengh of the i-th Entity according to the Auto-CM weights matrix;
   (This quantity will be used in equation 38b, below in the text).

$$Hid_i = \sum_j^N w_{i,j}; \qquad (4.68)$$

$$Hid_{max} = \max_i\{Hid_i\}; \qquad (4.69)$$

$$S_i = \frac{Hid_i}{Hid_{max}}. \qquad (4.70)$$

We now have to cope with two different metrics, namely the Euclidean one that defines the position of each entity within the geographical space, and the Auto-CM one which describes the quasi-distance of the same entities in a suitably re-mapped space. We need to merge the information built into these two matrices within a common framework. More precisely, we need to calculate how the Auto-CM matrix modifies the Euclidean space among the entities and how all the quantities that we have already introduced will be modified accordingly.

First of all, we must suitably re-write (4.6) and (4.11), ruled by the α parameter; (4.21) and (4.26), ruled by the β parameter; and, finally, (4.44) and (4.49), ruled by the γ parameter.

As to the equations that refer to α, we have:

$$p_i(\alpha) = \frac{1}{N-1} \sum_{j=1,j\neq i}^{N} e^{-\frac{d_{i,j}}{D}\alpha};\qquad(4.6)$$

and

$$p_i = \frac{1}{N-1} \sum_{j=1,j\neq i}^{N} e^{-\frac{d_{i,j}}{D}\alpha^*};\qquad(4.11)$$

change into:

$$p_i(\alpha) = \frac{1}{N-1} \sum_{j=1,j\neq i}^{N} e^{-\frac{d_{i,j}+d_{i,j}^{[AutoCM]}}{2\cdot D}\alpha};\qquad(4.6a)$$

and

$$p_i = \frac{1}{N-1} \sum_{j=1,j\neq i}^{N} e^{-\frac{d_{i,j}+d_{i,j}^{[AutoCM]}}{2\cdot D}\alpha^*};\qquad(4.11a)$$

As to β, we have:

$$p_i(\beta) = \frac{1}{N-1} \sum_{j=1,j\neq i}^{N} e^{-\frac{d_{i,j}}{D}\beta};\qquad(4.21)$$

and

$$p_i = \frac{1}{N-1} \sum_{j=1,j\neq i}^{N} e^{-\frac{d_{i,j}}{D}\beta^*};\qquad(4.26)$$

change into:

$$p_i(\beta) = \frac{1}{N-1} \sum_{j=1, j \neq i}^{N} e^{-\frac{d_{i,j}+d_{i,j}^{[AutoCM]}}{2 \cdot D}\beta};$$ (4.21a)

and

$$p_i = \frac{1}{N-1} \sum_{j=1, j \neq i}^{N} e^{-\frac{d_{i,j}+d_{i,j}^{[AutoCM]}}{2 \cdot D}\beta^*};$$ (4.26a)

Finally, for γ, we have:

$$p_{i,j}(\gamma) = e^{-\frac{d_{i,j}}{D}\gamma};$$ (4.44)

and

$$p_{i,j} = e^{-\frac{d_{i,j}}{D}\gamma^*};$$ (4.49)

change into

$$p_{i,j}(\gamma) = e^{-\frac{d_{i,j}+d_{i,j}^{[AutoCM]}}{2 \cdot D}\gamma};$$ (4.44a)

and

$$p_{i,j} = e^{-\frac{d_{i,j}+d_{i,j}^{[AutoCM]}}{2 \cdot D}\gamma^*};$$ (4.49a)

The re-mapping of coordinates as determined by the Auto-CM matrix also influences the calculation of the Proximity Scalar Field. Consequently, 4.38 has to be accordingly modified as well, by taking into account the strength of each entity in relation to the geographical points:

$$p_k = \frac{1}{N} \sum_{j=1}^{N} e^{-\frac{m_{k,j}}{D}\beta^*};$$ (4.38)

Equation 4.38 is changed into:

$$p_k = \frac{1}{N} \sum_{j=1}^{N} e^{-\frac{m_{k,j} \cdot q_{k,j}}{D}\beta^*};$$ (4.38a)

$$q_{k,j} = \sqrt{1.0 - \frac{S_j}{m_{k,j}}};$$ (4.38b)

for S_j see (4.70)

4.4.1   A Basic Example of Interaction Between Semantic Features and Physical Position

We start again from an 'artificial' example that allows us to elucidate some basic features of the methodology. Let us start by considering four entities distributed at the corners of a unit square:

Euclidean distances	P1	P2	P3	P4
P1	0.000000	1.000000	1.000000	1.414214
P2	1.000000	0.000000	1.414214	1.000000
P3	1.000000	1.414214	0.000000	1.000000
P4	1.414214	1.000000	1.000000	0.000000

Because of the symmetry, the TWC and the Centre of mass will stay in the same position (Fig. 4.26):

Consequently, the paths between the entities (see (4.44) will look like this Fig. 4.27):

The lines connecting the entities located along the diagonals appear smoothly warped. That happens because the diagonal of the unit square is an irrational number. The point is that the Relative TWC (4.38, 4.39, 4.40, 4.41, 4.42, 4.43, 4.44, 4.45, 4.46, 4.47, and 4.48) is sensitive to this numerical approximation. In fact, the scalar field of the trajectories draws out a sort of a 3-dimensional parabola, with some asymmetry (Fig. 4.28):

Now, we add to the previous matrix of distances a new matrix, generated by an AutoCM system that processes the similarities among the entities (similarity of shape, color, etc.). What we have to understand, then, is how the two matrices will actually interact. If the AutoCM matrix is identical to the first one, nothing will change. So we will consider the case in which this second matrix presents a very small difference with respect to the Euclidean one (1/100000):

AutoCM distances	P1	P2	P3	P4
P1	0.000000	**0.999990**	1.000000	1.414214
P2	1.000000	0.000000	1.414214	1.000000
P3	1.000000	1.414214	0.000000	1.000000
P4	1.414214	1.000000	1.000000	0.000000

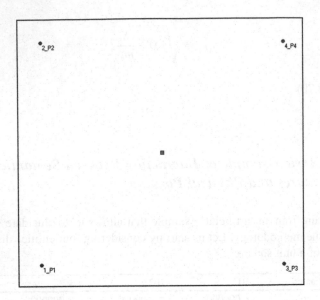

Fig. 4.26 Centre of mass and TWC for an artificial problem

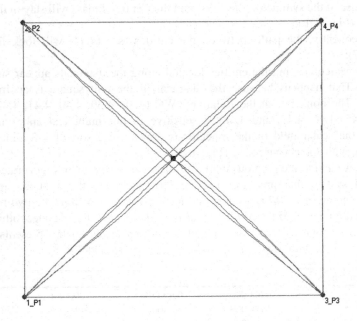

Fig. 4.27 Paths between entities

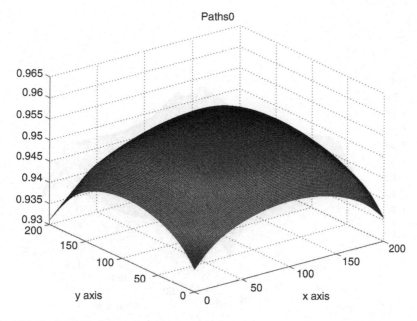

Fig. 4.28 Scalar field of the trajectories: symmetric case

In this new situation, the scalar field of the trajectories appears completely different (Fig. 4.29):

We then realize that a small perturbation that drives a wedge between the Euclidean metric and the Auto CM semi-metric causes the single-peak (quasi-) parabola to collapse into a complex surface with nine local maxima.

Now, we moreover modify the Auto CM matrix in three different, consistent ways (Figs. 4.30, 4.31, 4.32, 4.33 and 4.34).

Case 1. Trajectories
Case 1. Scalar Field of Trajectories
Case 2. Scalar Field of Trajectories

Case 1	Auto CM semi-metric				
P1 & P2	Distances	P1	P2	P3	P4
are closer	P1	0	**0.5**	1	1.414214
	P2	**0.5**	0	1.414214	1
	P3	1	1.414214	0	1
	P4	1.414214	1	1	0

Case 2	Auto CM semi-metric				
P1 & P2	Distances	P1	P2	P3	P4
Collapse in the	P1	0	**0**	1	1.414214
same point	P2	**0**	0	1.414214	1
	P3	1	1.414214	0	1
	P4	1.414214	1	1	0

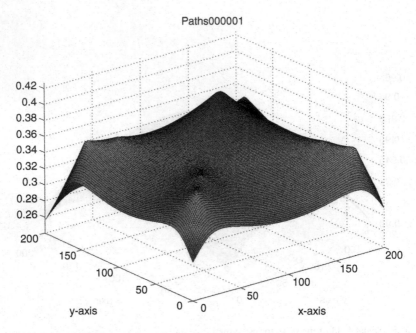

Fig. 4.29 Scalar field of the trajectories: perturbed case

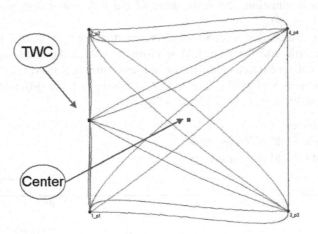

Fig. 4.30 Path between entities, case 1

Case 3: Trajectories:

Case 3	Auto CM semi-metric				
P1 & P2 & P3	Distances	P1	P2	P3	P4
are closer	P1	0	0.5	0.5	1.414214
	P2	0.5	0	1.414214	1
	P3	0.5	1.414214	0	1
	P4	1.414214	1	1	0

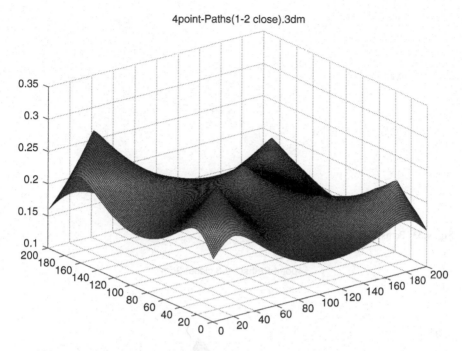

Fig. 4.31 Scalar field of trajectories, case 1

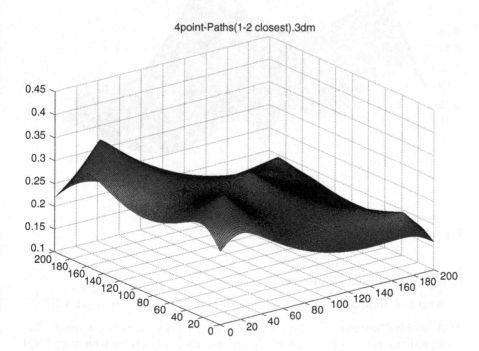

Fig. 4.32 Scalar field of trajectories, case 2

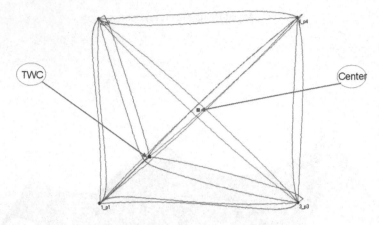

Fig. 4.33 Paths between entities, case 3

4point-Paths(1-2-3 close).3dm

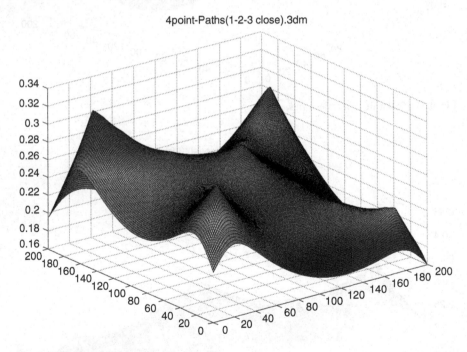

Fig. 4.34 Scalar field of trajectories, case 3

Case 3. Scalar Field of Trajectories

What kind of lessons can we draw from the above exercise? Here are a few:

1. A minimal distortion of symmetry in the distances generates a remarkable transition from a surface with only one maximum to a surface with many local maxima.

Fig. 4.35 6 × 6 grid

2. The resulting surface maintains a linear correlation with the new, non symmetric distance matrix.
3. The TWC tends to be located in the position with the highest density of close-by vertices, like an electro-magnetic field.

We have then repeated the same experiment with many points distributed along a regular, bi-dimensional grid (Fig. 4.35). We consider here the case of a 6 × 6 regular grid. In case 1, the Euclidean metric and the Auto CM distances are the same. Consequently, the surface of trajectories should be symmetric, but with numerical approximations (Fig. 4.36):

Case 1: Regular Grid 6×6
Case 1: Scalar Field of Trajectories when Auto CM Matrix = Euclidean Matrix

In case 2, we modified the Auto CM matrix as follows:

AutoCM_Matrix = Euclidean_Matrix + χ;
where χ = random_number $\in$ [0.0,0.1].

So, we perturb the Euclidean matrix by a very small amount of random noise. The shape of the scalar field of the trajectories changes considerably (Fig. 4.37):

Case 2: Scalar Field of Trajectories when Auto CM Matrix = Euclidean Matrix + Small Noise

36P-Paths(Regular).3dm

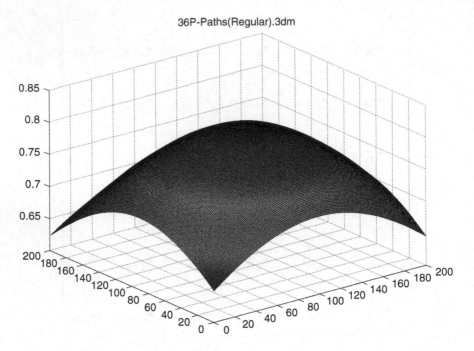

Fig. 4.36 Scalar field of trajectories, symmetric case

36-Paths(Light Rnd Noise).3dm

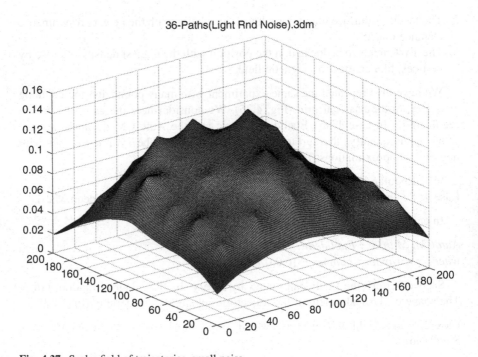

Fig. 4.37 Scalar field of trajectories, small noise

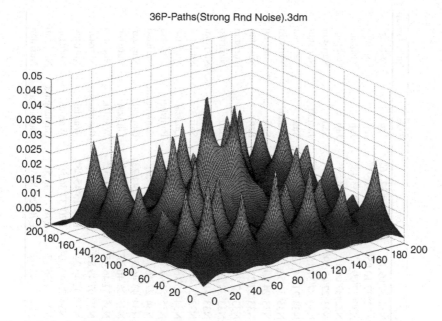

Fig. 4.38 Scalar field of trajectories, large noise

In case 3, we perturb the Auto CM matrix more intensely:

AutoCM_Matrix = Euclidean_Matrix + χ;
where χ = random_number ∈ [0.0,0.5].

That is to say, we add to the Euclidean matrix a substantially higher level of random noise. The shape of the scalar field of the trajectories now reveals a densely multi-peak surface (Fig. 4.38):

Case 3: Scalar Field of Trajectories when Auto CM Matrix = Euclidean Matrix + Large Noise

The examples of this subsection then illustrate quite clearly the intrinsic non-linearity of the TWC. Even small departures from symmetry or relatively small random perturbations cause a substantial differentiation of the TWC from the AC. This is a distinctive proof of the relevance of the semantics of space in determining, and reading, the deep structural properties of spatial phenomena.

Part D –Semantics of Physical Space: Applications

At this point we consider a small dataset about the terroristic attacks executed by terrorists in Afghanistan until May 2009.

This dataset consists of the latitude and longitudes of 50 attacks (Table 4.1 and Fig. 4.39). Each attack, further, is defined by 27 attributes, representing the estimation that USA forces did about the agent of the attack and its military equipment (Table 4.2).

Table 4.1 Lat and Long of 50 terrorist attacks in Afghanistan

Attack	Lat	Lon	Tribe	Ethnic Group	Attack	Lat	Lon	Tribe	Ethnic Group
1	N34°22'26,04"	E62°10'26,12"	Dari	Persiani	26	N32°37'09,19"	E65°52'32,16"	Jaji	Pashtun
2	N33°20'28,58"	E62°38'48,85"	Dari	Persiani	27	N32°37'10,05"	E65°52'32,91"	Jaji	Pashtun
3	N33°20'28,55"	E62°38'48,60"	Dari	Persiani	28	N32°37'09,10"	E65°52'33,48"	Jaji	Pashtun
4	N33°20'29,15"	E62°38'47,55"	Dari	Persiani	29	N32°37'09,12"	E65°52'33,34"	Jaji	Pashtun
5	N31°35'13,82"	E64°19'13,44"	Durrani	Pashtun	30	N33°27'03,59"	E69°44'26,19"	Safi	Nuristan
6	N32°21'16,22"	E63°23'58,74"	Durrani	Pashtun	31	N33°27'37,15"	E69°44'13,89"	Jadran	Nuristan
7	N31°35'13,40"	E64°19'33,84"	Durrani	Pashtun	32	N33°21'34,55"	E69°44'54,71"	Safi	Nuristan
8	N31°35'14,22"	E64°19'25,65"	Durrani	Pashtun	33	N33°31'53,21"	E69°44'56,26"	Jadran	Nuristan
9	N31°35'14,50"	E64°19'36,20"	Durrani	Pashtun	34	N33°24'30,11"	E69°45'32,43"	Safi	Nuristan
10	N31°35'13,58"	E64°19'03,31"	Durrani	Pashtun	35	N33°27'31,34"	E69°44'54,76"	Safi	Nuristan
11	N31°35'13,71"	E64°19'02,48"	Durrani	Pashtun	36	N33°27'36,51"	E69°43'43,39"	Safi	Nuristan
12	N31°35'13,16"	E64°1912,14"	Durrani	Pashtun	37	N33°34'23,98"	E69°44'33,87"	Safi	Nuristan
13	N31°35'13,33"	E64°19'43,27"	Durrani	Pashtun	38	N33°31'35,18"	E69°44'01,54"	Safi	Nuristan
14	N32°37'09,92"	E65°52'33,23"	Jaji	Pashtun	39	N33°22465,66"	E69°44'26,17"	Safi	Nuristan
15	N32°37'09,90"	E65°52'34,19"	Jaji	Pashtun	40	N34°15'46,27"	E70°49'05,29"	Ghilzai	Nuristan
16	N32°37'08,85"	E65°52'32,66"	Jaji	Pashtun	41	N35°51'46,71"	E71°38'16,76"	Panjsheri	Nuristan
17	N32°37'09,14"	E65°52'33,11"	Jaji	Pashtun	42	N34°22'46,16"	E71°55'43,87"	Samangani	Nuristan
18	N32°37'09,66"	E65°52'33,49"	Jaji	Pashtun	43	N34°43'46,31"	E70°08''44,23v	Ghilzai	Nuristan
19	N32°37'08,79"	E65°52'34,20"	Jadran	Pashtun	44	N35°25'46,77"	E70°53'46,59"	Ghilzai	Nuristan
20	N32°37'09,15"	E65°52'34,97"	Jaji	Pashtun	45	N35°37'46,65"	E70°02'51,32"	Wardak	Nuristan
21	N32°37'09,99"	E65°52'33,15"	Jaji	Pashtun	46	N35°07'49,81"	E72°41''33,66"	Panjsheri	Nuristan
22	N32°37'09,27"	E65°52'33,55"	Jaji	Pashtun	47	N34°02'46,24"	E70°15'45,88"	Wardak	Nuristan
23	N32°37'09,61"	E65°52'32,67"	Jadran	Pashtun	48	N34°28'47,60"	E70°37'36,59"	Wardak	Nuristan
24	N32°37'09,52"	E65°52'35,43"	Jadran	Pashtun	49	N34°44'56,27"	E70°47'45,53"	Samangani	Nuristan
25	N32°37'09,41"	E65°52'33,76"	Jadran	Pashtun	50	N34°35'43,07"	E70°29'06,69"	Samangani	Nuristan

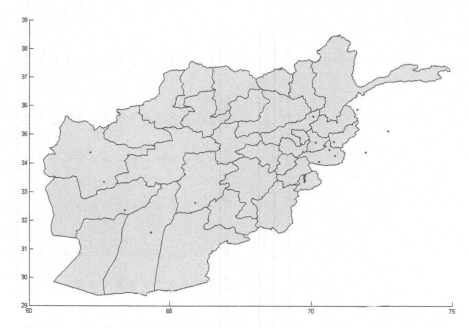

Fig. 4.39 The map of the 50 attacks. The x- and y-axes are latitude and longitude abbreviations

We have analyzed the dataset describing the attributes of the agents of each attack using AutoCM NN because we want to understand how these attacks are connected to each other and their natural clustering. Figure 4.40 shows the Maximally Regular Graph of the attacks, provided by AutoCM. A specific group of attacks seems to be the root of the others (attack: 14, 15, 16, 17, 18, 20, 21, 22, 26, 27, 28, and 29). All these attacks, except for attacks 19, 23, 24, and 25, have the same coordinates. Following the AutoCM inferences we are able to rebuild the connections and the interrelationships of the 50 attacks on the maps (Fig. 4.41) and to point out the main agent of these attacks (Table 4.3).

According this this view, the origin of attacks would be the region of Tarin-Kowt. From this location the other attacks would be expanded in the North-East and in the South-West. AutoCM infers, also, that the attacks in the Kabul area work as a strategic signal for the following attacks in the Herat area. This cognitive connection opens the possibility of a physical connection between the East and the West of Afghanistan from terrorists' point of view.

The independent application of TWC algorithm (4.1, 4.2, 4.3, 4.4, 4.5, 4.6, 4.7, 4.8, 4.9, 4.10, 4.11, 4.12, 4.13, 4.14, and 4.15), reinforced by the AutoCM metric (4.6a and 4.11a), locates the Hidden Point (TWC(alpha)) close to the Tarin city (look the Fig. 4.42). The same TWC, without AutoCM metric is less interesting, because it points the exact location of this group of twelve attacks in Tarin city.

According to the TWC algorithm, the Hidden Point is the point from which the distribution of the other points is managed, if we presuppose that all the points are managed by only one hand. Strategically speaking, the TWC locates the logistic

Table 4.2 Attributes of the tribes involved in the attacks

| Tribù | Etnia Group | Militanti | Armamenti Leggeri | | | Mezzi Pesanti | | Veicoli | | | Velivoli | | | Esplosivi Mine | | C4/TnT (ton) |
			Personali (Kalashnikov)	Antiaereo (Stinger)	Di gruppo (Mitragliere)	Carri	Blindati	Jeep	Camion	Moto	Aerei	Elicotteri	UAV	Anticarro	Antiuomo	
Durrani	Pashtun	1000	1500	50	30	10	5	25	50	100	1	1	3	1000	2500	2.0
Ghilzai	Pashtun	800	1000	50	50	3	10	30	30	60	0	1	0	500	800	0.5
Jadran	Pashtun	650	1000	0	10	5	3	20	25	10	0	0	0	350	200	1.0
Jaji	Pashtun	1200	2000	100	100	20	10	10	60	150	1	2	0	600	2000	3.5
Khugian	Pashtun	800	900	30	10	2	0	15	15	0	0	0	0	300	450	0.5
Mangal	Pashtun	900	1100	80	30	5	1	22	28	13	0	0	0	150	600	1.0
Mohamand	Pashtun	250	300	0	30	0	0	3	5	10	0	0	0	50	100	0.0
Safi	Pashtun	1300	1500	40	100	6	2	25	35	95	2	2	1	500	750	4.0
Shinwari	Pashtun	300	400	10	5	0	0	2	8	13	0	0	0	20	100	0.5
Tani	Pashtun	450	500	20	5	0	0	3	11	45	0	0	0	40	150	0.5
Wardak	Pashtun	350	400	10	10	1	2	4	7	15	0	0	0	100	150	0.0
Panjsheri	Tajiks	400	500	50	10	0	1	2	20	35	0	0	0	150	200	0.0
Andarabi	Tajiks	500	550	0	20	0	0	5	30	28	0	0	0	170	200	1.0
Samangani	Tajiks	350	400	10	20	0	0	1	14	12	0	0	0	30	60	0.0
Badakhshi	Tajiks	450	600	30	40	0	1	5	26	30	0	0	0	100	100	0.0

Tribe	Etnic Group	Militants	Light			Heavy		Vehicles			Vehicles			Mines		C4/TnT- (ton)
			Kalash-nikov	Stinger	Machine gun	Tank	Armored vehicles	Jeep	Truck	Motor-bike	Warplane	Elicopter	UAV	Anti_Tank_Bomb	Anti_Men_Bomb	
Durrani	Pashtun	1000	1500	50	30	10	5	25	50	100	1	1	3	1000	2500	2.0
Ghilzai	Pashtun	800	1000	50	50	3	10	30	30	60	0	1	0	500	800	0.5
Jadran	Pashtun	650	1000	0	10	5	3	20	25	10	0	0	0	350	200	1.0
Jaji	Pashtun	1200	2000	100	100	20	10	10	60	150	1	2	0	600	2000	3.5
Khugian	Pashtun	800	900	30	10	2	0	15	15	0	0	0	0	300	450	0.5
Mangal	Pashtun	900	1100	80	30	5	1	22	28	13	0	0	0	150	600	1.0
Mohamand	Pashtun	250	300	0	30	0	0	3	5	10	0	0	0	50	100	0.0
Safi	Pashtun	1300	1500	40	100	6	2	25	35	95	2	2	1	500	750	4.0
Shinwari	Pashtun	300	400	10	5	0	0	2	8	13	0	0	0	20	100	0.5
Tani	Pashtun	450	500	20	5	0	0	3	11	45	0	0	0	40	150	0.5
Wardak	Pashtun	350	400	10	10	1	2	4	7	15	0	0	0	100	150	0.0
Panjsheri	Tajiks	400	500	50	10	0	1	2	20	35	0	0	0	150	200	0.0
Andarabi	Tajiks	500	550	0	20	0	0	5	30	28	0	0	0	170	200	1.0
Samangani	Tajiks	350	400	10	20	0	0	1	14	12	0	0	0	30	60	0.0
Badakhshi	Tajiks	450	600	30	40	0	1	5	26	30	0	0	0	100	100	0.0

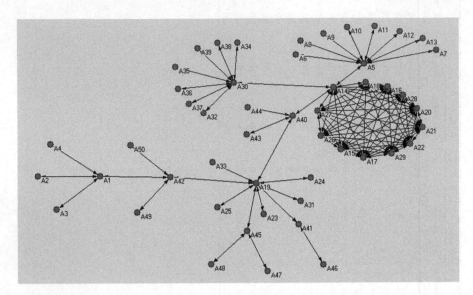

Fig. 4.40 Attacks logic networks according to AutoCM algorithm

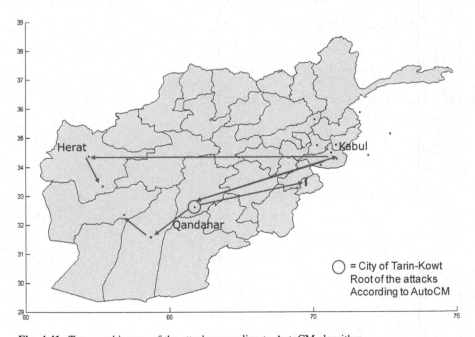

Fig. 4.41 Topographic map of the attacks according to AutoCM algorithm

Table 4.3 Jajl tribe is seemingly the main agent of this group of attacks

CM Types	A14	A15	A16	A17	A18	A20	A21	A22	A26	A27	A28	A29
tribe_Dari	-0.000003	-0.000003	-0.000003	-0.000003	-0.000003	-0.000003	-0.000003	-0.000003	-0.000003	-0.000003	-0.000003	-0.000003
tribe_Durrani	-0.000003	-0.000003	-0.000003	-0.000003	-0.000003	-0.000003	-0.000003	-0.000003	-0.000003	-0.000003	-0.000003	-0.000003
tribe_Ghilzai	-0.000003	-0.000003	-0.000003	-0.000003	-0.000003	-0.000003	-0.000003	-0.000003	-0.000003	-0.000003	-0.000003	-0.000003
tribe_Jadran	-0.000003	-0.000003	-0.000003	-0.000003	-0.000003	-0.000003	-0.000003	-0.000003	-0.000003	-0.000003	-0.000003	-0.000003
tribe_Jaji	0.990509	0.990509	0.990509	0.990509	0.990509	0.990509	0.990509	0.990509	0.990509	0.990509	0.990509	0.990509
tribe_Panjsheri	-0.000003	-0.000003	-0.000003	-0.000003	-0.000003	-0.000003	-0.000003	-0.000003	-0.000003	-0.000003	-0.000003	-0.000003
tribe_Safi	-0.000003	-0.000003	-0.000003	-0.000003	-0.000003	-0.000003	-0.000003	-0.000003	-0.000003	-0.000003	-0.000003	-0.000003
tribe_Samangani	-0.000003	-0.000003	-0.000003	-0.000003	-0.000003	-0.000003	-0.000003	-0.000003	-0.000003	-0.000003	-0.000003	-0.000003
tribe_Wardak	-0.000003	-0.000003	-0.000003	-0.000003	-0.000003	-0.000003	-0.000003	-0.000003	-0.000003	-0.000003	-0.000003	-0.000003
Nuristan	-0.000003	-0.000003	-0.000003	-0.000003	-0.000003	-0.000003	-0.000003	-0.000003	-0.000003	-0.000003	-0.000003	-0.000003
Pashtun	-0.000003	-0.000003	-0.000003	-0.000003	-0.000003	-0.000003	-0.000003	-0.000003	-0.000003	-0.000003	-0.000003	-0.000003
Persian	-0.000003	-0.000003	-0.000003	-0.000003	-0.000003	-0.000003	-0.000003	-0.000003	-0.000003	-0.000003	-0.000003	-0.000003

DATA	A14	A15	A16	A17	A18	A20	A21	A22	A26	A27	A28	A29
tribe_Dari	0	0	0	0	0	0	0	0	0	0	0	0
tribe_Durrani	0	0	0	0	0	0	0	0	0	0	0	0
tribe_Ghilzai	0	0	0	0	0	0	0	0	0	0	0	0
tribe_Jadran	0	0	0	0	0	0	0	0	0	0	0	0
tribe_Jaji	1	1	1	1	1	1	1	1	1	1	1	1
tribe_Panjsheri	0	0	0	0	0	0	0	0	0	0	0	0
tribe_Safi	0	0	0	0	0	0	0	0	0	0	0	0
tribe_Samangani	0	0	0	0	0	0	0	0	0	0	0	0
tribe_Wardak	0	0	0	0	0	0	0	0	0	0	0	0
Nuristan (Etnic Group)	0	0	0	0	0	0	0	0	0	0	0	0
Pashtun (Etnic Group)	0	0	0	0	0	0	0	0	0	0	0	0
Persian (Etnic Group)	0	0	0	0	0	0	0	0	0	0	0	0

132   M. Buscema et al.

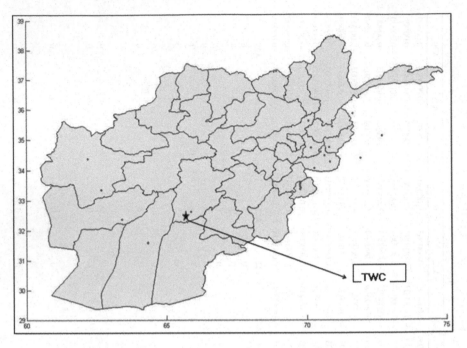

Fig. 4.42 The black star points the TWC(Alfa) generated by TWC algorithm

home of the attacks, close to the attacks analyzed by the AutoCM algorithm as the root for the other attacks, and close to the boundary between the big provinces of Uruzgan and Kandahar, exactly across the road joining Tarin-Kowt and Kandahar.

To understand the relevance of the semantic information, we compare the map of TWC(Beta) without the AutoCM metric (we call this the Syntactic Map), in Fig. 4.43a (see 4.16, 4.17, 4.18, 4.19, 4.20, 4.21, 4.22, 4.23, 4.24, 4.25, 4.26, 4.27, 4.28, 4.29, 4.30, 4.31, 4.32, 4.33, 4.34, 4.35, 4.36 and 4.37), and the same map produced using AutoCM metric in Fig. 4.43b (see 4.21a and 4.26a) called the Semantic Map. We remind the reader that the TWC(Beta) represents the map of probability of new points generation, from the analysis of the assigned points (the original 50 attack points). Consequently, the TWC(Beta) scalar field represents the vulnerability map of attacks, according to the stored data.

The difference between the two maps is self-evident. The Semantic Map is more extensive and tries to group all the attacks in a unique framework with two main clusters.

The TWC(Gamma) generates syntactic and semantic maps that are completely different. We further remind the reader that TWC(Gamma) builds its scalar field in consideration of the closeness of each point of the surface to the trajectories connecting each entity (attack location) to the others according to the constrains found by the scalar field of TWC(Beta).

The TWC(Gamma), conceptually speaking, is a very interesting quantity for it considers the possible trajectories connecting the place of the attacks, according to

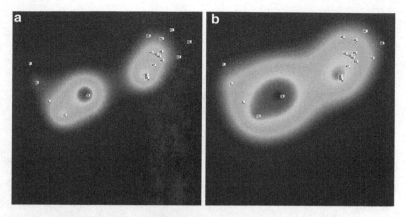

Fig. 4.43 (a) TWC(beta) Syntactic Map; (b) TWC(beta) Semantic Map

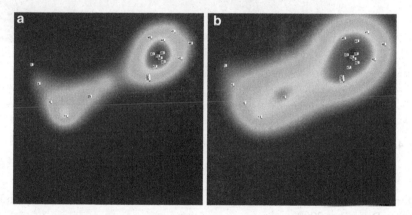

Fig. 4.44 (a) TWC(Gamma) Syntactic Map. (b) TWC(Gamma) Semantic Map

the TWC(Beta) scalar field then it implies a dynamic. Consequently, the TWC (Gamma) should figure out the scalar field of the *implicit consequences* of the 50 attacks thus generates a map of the risk of possible new attacks.

Also in this case we can generate a Syntactic Map (4.38, 4.39, 4.40, 4.41, 4.42, 4.43, 4.44, 4.45, 4.46, 4.47, 4.48, 4.49, 4.50, and 4.51) and a Semantic map (4.38a, 4.38b, 4.44a and 4.49a). Figure 4.44a and b show the Syntactic and the Semantic scalar field of the TWC(Gamma).

In this case the Semantic Map better represents the general framework of the attacks and their connections. But there is something new in both the TWC (Gamma) maps: the area of the new possible attacks has moved to the North-East of Afghanistan, in the Kabul province and in Pakistan. That is really surprising for the 50 attacks of the assigned dataset were updated to May of 2009, a real explosion

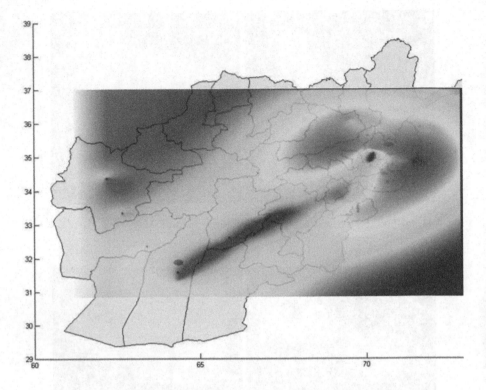

Fig. 4.45 Gradient of the TWC(Gamma), Semantic Map (that is, using AutoCM metric)

of terroristic attacks occurred exactly at north and at east of Kabul from September 2009 to December 2009.

This predictive capability is dependent on the data: TWC(Gamma) is able to extract from the data most of their subtle consequences, and many of them are statistically meaningful. Moreover, the more a situation continues, the fewer choices remain available.

The calculation of the gradient of the TWC(Gamma) scalar field (see Sect. 2.4), shows a possible future map of the instability in Afghanistan; in what areas is it more plausible to have terroristic attacks in highly unpredictable ways (Fig. 4.45)?

In Fig. 4.45 the areas around Herat and Ferat also shows a relevant probability of being subjected to terroristic attacks after May 2009...and that is exactly what has happen.

We need to conclude this exemplary application of the TWC algorithm showing the Non Linear MSTs generated with the support of AutoCM metric (the Semantic method) and without AutoCM NN, that is, using only the Euclidean distances (the Syntactic method).

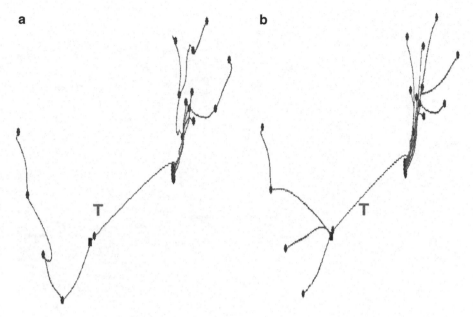

Fig. 4.46 (a) Syntactic NL-MST. (b) Semantic NL_MST

Figure 4.46a shows the Syntactic NL-MST, and Fig. 4.46b show the Semantic NL-MST of the 50 attacks.

Comparing the two NL-MSTs with the map of Afghanistan (Fig. 4.45) is impressive how much both the trajectories match with the road networks of the Region. But there is a meaningful difference: while the syntactic NL-MST tends to represent the connections among the attack locations by, following the only highway that crosses the south of Afghanistan, the semantic NL-MST chooses the minor roads connecting the same places, and that seems to have a greater likelihood.

The second meaningful difference is concerned with the semantic NL-MST map of the Kabul area that suggests that it is to become the new center of the attacks. The trajectories are very compact and they tend to work as a road-ring between the north and the east of the region.

If now we consider that we started from the Latitude and the Longitude of 50 attacks and 27 variables, we need to note that TWC and AutoCM algorithms are suitable for an intensive data mining examination and analysis. They have shown to be able to extract usefully valuable information from the "bottom of the basket."

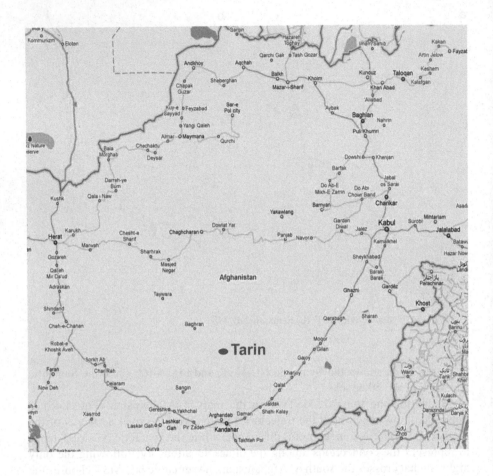

4.5 Concluding Remarks

Our analysis shows how a relatively straightforward mathematics may help decision makers in situations characterized by a substantially limited amount of information, and moreover how the mathematics of complex systems can improve the level of accuracy in the analysis of spatial phenomena, compared with the level that can be achieved with a classical statistical toolbox. We have illustrated a few applications that deliver interesting and encouraging results, but we are at the same time aware that the methodology presented here needs more, careful verification and validation on further logically consistent and empirically relevant problems.

Models concerning the onset and diffusion of epidemics, for example, describe the spread of infectious diseases across populations. More and more, these models are being used for predicting, understanding and developing control strategies.

In realistic epidemic models, a key issue to consider is the representation of the interaction process through which the disease spreads over, and network-based models have qualified as viable candidates, not only at the micro- but also at the macro-level, predicting which cities are endemic sources for the disease, and understanding the shape and dynamics of diffusion waves may become fundamental to design optimal surveillance and control strategies. In particular, our results are consistent with the idea that the spreading of an infectious disease is not random but follows a geometry which reflects inherent, as yet undiscovered mathematical laws based on some probabilistic density function.

Likewise, there may be many other social phenomena that could be characterized by the same fundamental structure and level of complexity. At present, we cannot count on a methodology that allows us to tackle this sort of issues adequately, but we will pursue this objective in future work. We are convinced that these are but a few examples of the actual range of problems from many different disciplines that may be usefully addressed by means of the TWC methodology. We look forward to future research exploring new ground in this respect, and providing an expanding literature on which to build a powerful and consistent theory of complex spatial dynamics.

References

Brantingham PL, Brantingham PJ (1981) Environmental criminology. Waveland Press Inc., Prospect Heights, Illinois

Brantingham PL, Brantingham PJ (1984) Patterns in crime. Macmillan, New York

Buscema M (2008–2012) 16 Meta Auto-Associative, Semeion Software #51. Semeion, Rome

Buscema M (2004) Genetic doping algorithm (GenD). Theory and applications. Expert Syst 21 (2):63–79

Buscema M (2007a) Squashing theory and contractive map network. Semeion Technical Paper #32, Rome

Buscema M (2007b) PST with Hidden Unit Generation, version 8.0. Semeion Software #11, Rome, 1999–2007

Buscema M (2007c) A novel adapting mapping method for emergent properties discovery in data bases: experience in medical field. In: 2007 I.E. international conference on systems, man and cybernetics (SMC 2007), Montreal, Canada, pp 7–10

Buscema M (2008) PST Cluster and TWC, version 6.0. Semeion Software #34, Rome, 2006–2008

Buscema M, Grossi E (2008) The semantic connectivity map: an adapting self-organizing knowledge discovery method in data bases. Experience in gastro-oesophageal reflux disease. Int J Data Min Bioinform 2:362–404

Buscema M, Grossi E, Breda M, Jefferson T (2009a) Outbreaks Source: A New Mathematical Approach to Identify Their Possible Location. Physica A 388:4736–4762

Buscema M, Grossi E, Snowdon D, Antuono P (2008a) Auto-contractive maps: an artificial adaptive system for data mining: an application to Alzheimer disease. Curr Alzheimer Res 5:481–498

Buscema M, Helgason C, Grossi E (2008b) Auto contractive maps, H function and maximally regular graph: theory and applications, special session on artificial adaptive systems in medicine: applications to the real world. NAFIPS 2008 (IEEE), New York

Buscema M, Monaghan G, Richards P (eds) (2007) CDTD, Central Drug Trafficking Database, in MPS Drugs Strategy 2007–2010 & Delivery Plan, Appendix 1, Chapter 6, London Metropolitan Police, London

Buscema M, Sacco PL (2008) Auto-contractive maps, the H function and the maximally regular graph (mrg): a new methodology for data mining, Mimeo, Semeion, Rome

Buscema M, Sacco PL, Terzi S (2009b) Take five: Popperian openness and the emergent world order, under review, 2009

Buscema M, Sacco P (2011) An Artificial Intelligent Systems Approach to Unscrambling Power Networks in Italy's Business Environment, in Capecchi et al (eds), Applications of Mathematics in Models, Artificial Neural Networks and Arts, Springer, Netherland

Buscema M, Terzi S (2006) A new evolutionary approach to topographic mapping. Proceedings of the 7th WSEAS international conference on evolutionary computing, Cavtat, Croatia, June 12–14, 2006, pp 12–19

Buscema M, Terzi S (2006a) PST: an evolutionary approach to the problem of multi dimensional scaling. WSEAS Trans Inf Sci Appl 3(9):1704–1710

Canter D, Larkin P (1993) The environmental range of serial rapists. Journal of Environmental Psychology 13:63–69

Canter D, Tagg S (1975) Distance estimation in cities. Environ Behav 7:59–80

Canter D (1999) Modelling the home location of serial offenders. Paper presented at the 3rd annual international crime mapping research conference, Orlando, December 1999

Canter D (2003) Mapping murder: the secrets of geographic profiling. Virgin Publishing, London

Canter D, Coffey T, Huntley M, Missen C (2000) Predicting serial killer s' home base using a decision support system. J Quant Criminol 16:457–478

Cliff AD, Haggett P (1988) Atlas of disease distributions: analytic approaches to epidemiologic data. Oxford University Press, Oxford

Coelho FC, Cruz OG, Codeco CT (2008 Feb 26) Epigrass: a tool to study disease spread in complex networks. Source Code Biol Med 3(1):3

Cromley EK, McLafferty SL (2002) GIS and public health. The Guilford Press, New York, pp 189–209

Duda RO, Hart PE (1973) Pattern classification and scene analysis. Wiley, New York, pp 271–272

Eng SB, Werker DH, King AS, Marion SA, Bell A, Issac-Renton JL, Irwin GS, Bowie WB (1999) Computer-generated dot maps as an EpidemiologicTool: investigating an outbreak of toxoplasmosis. Emerg Infect Dis 5:815–819

Geli P, Rolfhamre P, Almeida J, Ekdahl K (2006) Modelling pneumococcal resistance to penicillin in southern Sweden using artificial neural networks. Microb Drug Resist 12(3):149–57

Jaehne B, Scharr H, Koerkel S (1999) Principles of filter design. In: Jaehne B, Haussecker H, Geissler P (eds) Handbook of computer vision and applications, vol 2. Academic Press, New York, pp 125–151

Kim M, Choi CY, Gerba CP (2007) Source tracking of microbial intrusion in water systems using artificial neural networks. Water Res Oct 10 [Epub ahead of print]

Kohonen T (1995) Self-organizing maps. Springer Verlag, Berlin

Levine N and Associates (2004) CrimeStat III – a spacial statistical program for the analysis of crime incident locations. Ch 10, pp 10.1–10.2. The National Institute of Justice, Washington DC

Linroth K (1890) Influensan i Sverige 1889–1890 enligt iakttagelser af landets läkäre. Del. I: Influensan i epidemiologiskt hänseende. Svenska Läkaresällskapets Nya Handlingar. Serie III, 1–92, 1890

O'Leary M (2006) A new mathematical technique for geographic profiling. The NIJ Conference, Washington DC

Parham PE, Ferguson NM (2006) Space and contact networks: capturing the locality of disease transmission. J R Soc Interf 9483–9493

Pyle GF (1979) Studies of disease diffusion: applied medical geography. V.H. Winston & Sons, Washington, DC, p 123

Rich T, Shively M (2004) A methodology to evaluating geographic profiling software, Document No. 208993, Award Number ASP T-037, by Abt Associate Inc for NIJ

Rossmo DK (1993) Target patterns of serial murderers: a methodological model. Am J Criminal Justice 17:1–21

Rossmo DK (1995) Overview: multivariate spatial profiles as a tool in crime investigation. In: Rebecca Block C, Dabdoub M, Fregly S (eds) Crime analysis through computer mapping. Police Executive Research Forum, Washington, DC, pp 65–97

Rossmo DK (1997) Geographic profiling. In: Janet LJ, Debra A (eds) Bekerian, offender profiling: theory, research and practice. John Wiley and Sons, Chichester, pp 159–175

Rossmo DK (2000) Geographic profiling. CRC Press, Boca Raton, FL

Rossmo DK (2005) An evaluation of NIJ's evaluation methodology for geographic profiling software. Available on the MAPS website

Rytokonen MJ (2004a) Not all maps are equal: GIS and spatial analysis in epidemiology. Int J Circumpolar Health 63(1):18

Rytokonen MJP (2004b) Not all maps are equal: GIS and spatial analysis in epidemiology. Int J Circumpolar Health 63(1):11

Skog L, Hauska H, Linde A (2008) The Russian Influenza in Sweden 1889–1890: an example of geographical information system analysis. Eurosurveillance 13:1–7

Snow J (1855) On the mode of communication of cholera, 2nd edn. Churchill, London, 1853

Walter SD (2000) Disease mapping: a historical perspective. In: Elliot P, Wakfield JC, Best NG, Briggs DJ (eds) Spatial epidemiology: methods and applications. Oxford University Press, Oxford, p 225

Chapter 5
Meta Net: A New Meta-Classifier Family

Massimo Buscema, William J. Tastle, and Stefano Terzi

5.1 Introduction

The purpose of a classification system is to perform the task of categorization on some object and to do so with a reasonable degree of accuracy. There exists today a rather extensive listing of classifiers developed around specialized algorithms to satisfy certain classification schemes. This has led to the creation of a vast library of available instruments from which an investigator must make a choice, with each classifier possessing a particular typology. While one type of classifier might yield excellent results in one situation, it might also yield dismal results when applied to another.

What has become conspicuously apparent from the creation of these many classifiers over an extended time period is that a perfect "classifier" does not exist. Even if it were possible to create some sort of hierarchy with respect to the efficacy of each typology of classifier, the evidence suggests that the results make sense when applied to the set of test data utilized to train and evaluate the investigated model but when given over to data derived from a "real world" problem, different classifiers typically identify different solutions. All classifiers engage in the standard process of training with some appropriate validation protocol, but the task at hand is to select, from the different typologies of classifiers, the one with the best characteristics possible. If we consider the classification process to be an exercise in data mining,

M. Buscema (✉)
Semeion Research Center of Sciences of Communication, Via Sersale 117, Rome, Italy

Department of Mathematical and Statistical Sciences, CCMB, University of Colorado,
Denver, Colorado, USA
e-mail: m.buscema@semeion.it

W.J. Tastle
Ithaca College, New York, USA

S. Terzi
Semeion Research Center of Sciences of Communication, Via Sersale 117, Rome, Italy

W.J. Tastle (ed.), *Data Mining Applications Using Artificial Adaptive Systems*,
DOI 10.1007/978-1-4614-4223-3_5, © Springer Science+Business Media New York 2013

we discover that each classifier can classify the same inputs into different classes. This means that as the quantity and quality of extracted information changes from classifier to classifier; some typologies of classifiers like neural networks and decision trees present a great internal variability, producing sensible but different models when applied to the same problem.

In the standard process we have briefly described above, the diversity of models that have been produced is exploited; one strategy is to choose a single classifier, excluding all the others. Another kind of strategy consists of mixing a subset of classifiers to exploit the possible complementarities of information that can be extracted from each classifier. Such a strategy of using the outputs of several classifiers to produce a combined, and improved, output is produced by what is called a meta-classifier.

Dietterich suggests three motivations to explain why the fusion of the single classifier should produce a more efficient one: one is a *statistic motivation*, the second is a *computational motivation*, and the third is a *representational motivation*. For a thorough review of these motivations the reader is directed to Dietterich (2002).

The problem of the construction of a meta-classifier is quite complex in that it requires a formal schematization of possible options with definitions of a terminology and taxonomy. Additionally, it also takes into account that due to the nature of different problems, this schema may well be subjected to various exceptions.

Kuncheva (2004) considers four possible dimensions of projects during the development of a meta-classifier:

1. The database for the training and the validation
2. The selection of the "significant" variables;
3. The choice and training of the single classifiers; and
4. The definition of a combination strategy.

The first three dimensions represent a forward propagation chain of variations which generates the final classifier:

- If two classifiers are trained with two different training sets, the two classifiers will develop different data models; [*boosting* is a machine learning meta-algorithm for performing supervised learning and is based on the premise that a set of weak learners might create a single strong learner; *bagging* is another machine learning algorithm for a classification based on the model averaging approach];
- If two classifiers are trained with the same records, but possessing different variables, the two classifiers will develop different data models;
- If two classifiers are trained with the same training set, but the mathematics of the two classifiers is different (topology, learning rule, signal dynamics or cost function), the two classifiers will develop different data models;
- If two classifiers are trained with the same training set and have the same algorithm, but begin the learning session with either initial random weights or parameters, then the two classifiers will develop different data models.

In any case, we direct our attention to dimension four above and focus on the development of particular strategies of combination of single classifiers. It must be pointed out that dimensions three and four are often strictly connected. The choices made with respect to the mathematics of the classifiers and the initial weights or parameters will have consequences on the possible choices for the other, and vice versa.

While many taxonomies of classification are available in literature, one good review source is a book by Kuncheva (2004) in which some keys are listed that can be used to organize ensembles of classifiers. However these typologies are very often a simple list of features. Instead, we intend to present a generative typology, able to underline relevant features of meta-classifiers such that it is possible to generate new algorithms.

The key features we have identified are:

Algorithmic **category** – a meta-classifier can define its characteristics in two ways:

> *Static* – a calculation of characteristics and results of composing classifiers executed in a non-iterative way. The algorithm does not plan an iterative analysis of composing characteristics of classifiers in order to define the best way of establishing its parameters. A static algorithm can be:
>
> > Flexible – a *vector* of parameters emerges from a calculation.
> > Strict – only *one* parameter emerges from a calculation.
>
> *Dynamic* – an iterative calculation based on characteristics and results in the composition of classifiers to optimize a vector of parameters. A dynamic algorithm can be:
>
> > Trainable – the iterative algorithm tends to define the data entry continuous function parameters.
> > Optimizable – the iterative algorithm tends to optimize some cost function.

Extensional **category** *(Scope)* – a meta-classifier can define its characteristics as:

> *Local* – each composing classifier, in an independent way, provides the meta-classifier with some characteristics.
> *Global* – characteristics and results of all composing classifiers interact, globally defining the meta-classifier characteristics.

Teleological **category** – a meta-classifier can define its characteristics as:

> *Supervised* – the relevance of each composing classifier is weighted on the basis of the right/wrong results it produces.
> *Autopoietic* – the relevance of each composing classifier is weighted on the basis of the produced results without considering mistakes or successes. Autopoietic meta-classifiers, obviously, offer interesting performances when all composing classifiers have a confusion matrix which respects the following condition:

$$\text{Target}_i = Err_{i,i} - \sum_{j=1, j \neq i}^{N} Err_{i,j} < 0;$$

Functional category – a meta-classifier can evaluate each new entry input in this manner:

> *Feed-forward* – a meta-classifier provides only one response for each new entry input;
>
> *Recursive* – a meta-classifier generates more responses, each considering the previous ones, until the process optimizes a specific cost function (providing always the same classification response). During the recall process, this kind of meta-classifier works as a dynamic system; when a new input is presented, each one of its components hypothesizes a class for it and then all components negotiate their different hypothesis until they dynamically reach an agreement. We have found no information on meta-classifiers of this kind in literature. At a future time we will introduce a meta-classifier possessing these features.

5.2   Proposed Algorithm: Meta-Net Metaclassifiers

5.2.1   General Properties

The fundamental characteristic of the Meta-Net (Kohavi and Provost 1998) consists of considering not only the "positive credibility" of its composing classifiers (i.e., "this pattern is white"), but also their "negative credibility" (i.e., "this pattern is not white"). So, the characterizing connection of the Meta-Net is to connect each output

METANET Topology

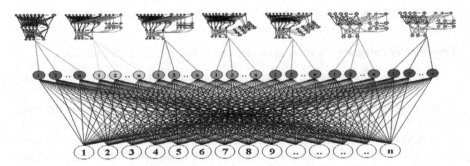

Fig. 5.1 Meta-Net general topology

node of each composing classifier with each output class. "Complete grid" connections are planned between Meta-Net inputs and outputs, and each connection can be either *excitatory* (positive numbers), or *inhibitory* (negative numbers). See Fig. 5.1.

Between 1994 and 2008 Semeion researchers conceived and developed a series of Meta-Classifiers (Buscema 1998e) based on some common traits and called them "Meta-Nets." All Meta-Nets have typically similar neural network architecture (Kohavi and Provost 1998); certain input nodes are the whole outputs of all composing classifiers, and certain output nodes are the output classes of the classification problem. The connections between Meta-Net inputs and outputs always possess a complete grid structure and are defined by specific algorithms characterizing the Meta-Net peculiarities.

The Meta-Net output vector is calculated in this way:

$$Class_i = \sum_j^N \sum_k^M I_j^k \cdot w_{i,j}^k;$$
$$Winner\_Class = \underset{i}{Arg\ Max}\ \{Class_i\}.$$

where:

N is the dimension of the confusion matrix;
M is the number of classifiers;
$w_{i,j}^k$ is the value of the weight connecting output node j of the kth
   base classifier to output node i of Meta-Net;
I_j^k is the output node j of the kth base classifier;
$Class_i$ is the output class of the Meta-Net;
$Winner\_Class$ is the winner class selected by Meta-Net.

All Meta-Nets are unsupervised. Each one evaluates its own output without knowledge of the errors in the composing classifiers; it only knows the statistic of their responses. So, Meta-Nets are *strongly sensitive to the quality of classifiers to be optimized*. This means that each Meta-Net, to be considered excellent, should be composed of classifiers in which the confusion matrix, in blind testing, clearly respects the following equation:

$$\forall k, k \in P : a_{i,i}^k - \sum_{j\neq i}^{N-1} a_{i,j}^k > 0;$$

where $a_{i,j}^k$ is a generic cell of the confusion matrix. However, in the tests that follow we shall verify that this condition, if not properly respected, will produce a "very smooth" fall of Meta-Net capacities in accordance with the typical characteristics of artificial neural networks (ANNs).

Each connection value represents the plausibility trough in which every component classifier supports every classification node of the Meta-Net. The numerical value of each Meta-Net connection can belong to the interval between $-\infty$

(implausibility) and $+\infty$ (plausibility). The plausibility and the implausibility of each connection is a function of the probability of each Meta-Net component during the testing phase.

5.2.2 Weight Definition

Weights are estimations based on the performance of base classifiers evaluated on an independent testing set. The results are summarized and used in the confusion matrix.

From a mathematical perspective, the common feature of the all Meta-Net algorithms is the specific procedure through which the plausibility of each output of any classifier is connected to each output of the global Meta-Classifier. To explain this procedure we need to start from the analysis of the confusion matrix of one classifier (Kohavi and Provost 1998):

$$
\textit{Classifier} \text{ k} \qquad \text{Output}
$$

$$
\text{Target} \qquad \begin{pmatrix} a_{11} & \cdots & a_{1N} \\ \cdots & \cdots & \cdots \\ a_{N1} & \cdots & a_{NN} \end{pmatrix}
$$

In this matrix we need to distinguish four criteria for each cell, $a_{i,j}^k$. The first criterion represents the "Rights," that is, the plausibility by which the kth classifier considers correct the records classified in the cell $v_{i,j}^k$ with respect to the summation of Targets:

$$
R_{i,j}^k = \frac{a_{i,j}^k}{\sum\limits_{j}^{N} a_{i,j}^k}
$$

The second criteria represents the "Corrects," that is, the plausibility by which the kth classifier considers "correct" the records classified in the cell $v_{i,j}^k$ with respect to the (column) summation of outputs:

$$
C_{i,j}^k = \frac{a_{i,j}^k}{\sum\limits_{i}^{N} a_{i,j}^k}
$$

The third criteria is a correlation of the "Rights" to the probability that any specific output depends on a specific Target: $p_{j,i}^k = p(O_j^k | T_i^k)$.

The fourth criteria is a correlation of the "Corrects" to the probability that any specific Target comes from a specific Output: $p_{i,j}^k = p(T_i^k | O_j^k)$.

Every weight connecting the output of each base classifier (that is, the Meta-Net input) and the output of the Meta-Net depends not only by the **sensitivity** of the considered classifier, but also by its **precision**. In other words, each weight of Meta-Net is the result of a function composed by the sensitivity and by the precision of each cell of the confusion matrix generated in test phase for each base classifier.

$$w_{i,j}^k = f\left(R_{i,j}^k, C_{i,j}^k\right).$$

Legenda:

$R_{i,j}^k$ = sensitivity of the cell i,j in the k-th basic classifier;
$C_{i,j}^k$ = precision of the cell i,j in the k-th basic classifier;
$f()$ = typically a fuzzy function;
$w_{i,j}^k$ = value of the weight between the j-th output of the k-th
classifier and the i-th output of Meta Net.

The function composing the sensitivity and the precision of each weight of a Meta Net can be a simple fuzzy rule, like the following:

$$w_{i,j}^k = \min\left\{R_{i,j}^k, C_{i,j}^k\right\}.$$

Or a more complex fuzzy rule like this one:

$$w_{i,j}^k = \left(R_{i,j}^k + C_{i,j}^k\right) - \left((1 - R_{i,j}^{\ k}) \cdot (1 - C_{i,j}^{\ k})\right).$$

Both the R and C matrices give additional information to the Meta-Classifier for the purpose of the weighting of each base classifier. The intention is to provide increased accuracy to the Meta-Classifier. Traditionally, the combination of the outputs from the base classifiers has been done with weighted averages and these weights have been determined by the main diagonal of R and C. By limiting the weight calculations to the diagonal omits potentially important additional information and hence, the precision of a value does not necessarily indicate conciseness of accuracy, and here is where the Meta-Classifier gains its value. It utilizes all the information available in the entire matrix to determine the weights of the Meta-Classifier.

Referring to Kuncheva's (2004) work, given L number of classifiers and c number of classes, we can have three types of weighted averages depending on the number of weights. First, we can have L weights in which each classifier has exactly one weight; second, we can have L * c weights in which there is one weight per class, and third we can have L * c * c weights which represent a complete connection between the outputs of the base classifiers and the outputs of the Meta-Classifier. The Meta-Net algorithm uses this third method to take into account the possibility of how much a single base classifier might render a wrong decision.

It is important to understand the meaning of the R and C values that are off the main diagonal. For the R matrix the values represent the number of times the base classifier answered "i" when the answer should have been "j", and the C matrix is the "precision" of the confusion between "i" and "j"; simply stated, C informs us that from among all the times the classifier answered class "i" (correct and incorrect decisions) the percentage of correct decisions was actually class "j".

In other words Meta Net algorithms additionally consider the **inhibitory credibility** of each base classifier. This is the case when the weight pushes Meta Net to change opinion in relation to the classification suggested by the base classifier. An example: Suppose the base classifier confuses the correct class A with the incorrect class B 30 times over 100. Let us also suppose that the base classifier makes systematic mistakes, confusing class A with class B. At this point the weight connecting class B of the base classifier with class A (the correct one) of Meta-Net will be strong, while the weight connecting class B of the base classifier with class B of Meta-Net will be weak. And, consequently, Meta-Net is also able to correct many systematic errors of classification generated by its base classifiers.

Armed with this theory we can now proceed to a description of the equations.

5.2.3   Specific Weight Equation on the Confusion Matrix

5.2.3.1   Meta-Bayes

Meta-Bayes is a meta-classifier created in 1994 by M. Buscema at Semeion and has been developed and refined through numeric testing until 2007 at which time was the progenitor and inspiration of all the family of meta-classifiers created at Semeion and known by the collective name Meta-Net.

A first version appeared in Buscema (1998e).

The specific weight equation is:

$$w_{i,j}^k = -\ln\left(\frac{(1 - R_{i,j}{}^k) \cdot (1 - C_{i,j}{}^k)}{R_{i,j}{}^k \cdot C_{i,j}{}^k}\right) \tag{5.1}$$

This equation is inspired to the "co-occurrence equation" used in precedent research by Rumelhart et al. (1986b) and derived from a Bayesian analysis of the probability that the unit x(i) should be on given unit x(j) and vice versa:

$$w_{i,j} = -\ln\left(\frac{p(x_i = 0 \,\&\, x_j = 1) \cdot p(x_i = 1 \,\&\, x_j = 0)}{p(x_i = 1 \,\&\, x_j = 1) \cdot p(x_i = 0 \,\&\, x_j = 0)}\right).$$

5.2.3.2 Meta-Sum

Meta-Sum is the name of an equation developed by M. Buscema in 2007 to optimize the weights matrix of a new Meta-Net:

$$w_{i,j}^k = \frac{1}{2} \cdot \left[\left(R_{i,j}^k + C_{i,j}^k \right) - \left((1 - R_{i,j}^k) \cdot (1 - C_{i,j}^k) \right) \right]. \tag{5.2}$$

In terms of fuzzy sets, if we pose $\tilde{A} = R_{i,j}^k$ and $\tilde{B} = C_{j,i}^k$ then we can write:

$$w_{i,j}^k = \frac{\mu_{\tilde{A}}(x^k) + \mu_{\tilde{B}}(x^k) - \left(1 - \mu_{\tilde{A}}(x^k) \right) \cdot \left(1 - \mu_{\tilde{B}}(x^k) \right))}{2}. \tag{5.3}$$

To better understand this equation we can proceed in this way:
If we pose:

$$x = \sum_{i}^{N} a_{i,j}^k; \, y = \sum_{j}^{N} a_{i,j}^k; \quad z = a_{i,j}^k.$$

then we can write (5.3) as:

$$w_{i,j} = \frac{1}{2} \cdot \left(\frac{z}{x} + \frac{z}{y} - \frac{x-z}{x} \cdot \frac{y-z}{y} \right);$$

and then:

$$w_{i,j} = \frac{1}{2} \cdot \left[\frac{z(x+y)}{xy} - \frac{xy - z(x+y) - z^2}{xy} \right] = \frac{2z(x+y) - xy - z^2}{2xy}.$$

5.2.3.3 Meta-Fuzzy

Meta-Fuzzy is a meta-classifier created in 2008 by M. Buscema at Semeion. Meta-Fuzzy can be considered the most simple member of the Meta-Net family. From a typological point of view Meta-Fuzzy meta-classifier is definable as follows:

$$w_{i,j}^k = \max \left\{ \min \left(R_{i,j}^k, F_{i,j}^k \right), \min \left(C_{i,j}^k, M_{i,j}^k \right) \right\} = \min \left\{ R_{i,j}^k, C_{i,j}^k \right\}.$$

In terms of fuzzy sets, if we pose $\tilde{A} = R_{i,j}^k$ and $\tilde{B} = C_{i,j}^k$ then we can write:

$$w_{i,j}^k = \max \{ \min (\mu_{\tilde{A}}(x^k), 1 - \mu_{\tilde{B}}(x^k)), \min (\mu_{\tilde{B}}(x^k), 1 - \mu_{\tilde{A}}(x^k)) \}$$
$$= \min \{ \mu_{\tilde{A}}(x^k), \mu_{\tilde{B}}(x^k) \}.$$

5.2.3.4 Meta-Exp

Meta-Exp is a meta-classifier created in 2008 by M. Buscema at Semeion. Meta-Exp can be considered an alternative version of Meta-Sum:

$$w_{i,j}^k = \frac{e^{\left(R_{i,j}{}^k + C_{i,j}{}^k\right)}}{e^{\left(M_{i,j}{}^k \cdot F_{i,j}{}^k\right)}};$$

(5.4)

5.2.3.5 Meta-Einstein

$$w_{i,j}^k = \frac{R_{i,j}^k + C_{i,j}^k}{1 + M_{i,j}{}^k \cdot F_{i,j}{}^k}$$

(5.5)

The inspiration for this equation is the Einstein Sum in Fuzzy Theory of Sets (Zimmermann 1996), with a little change in the denominator:

$$w_{i,j}^k = \frac{\mu_{\tilde{A}}\left(x^k\right) + \mu_{\tilde{B}}\left(x^k\right)}{1 + \left(1 - \mu_{\tilde{A}}(x^k)\right) \cdot \left(1 - \mu_{\tilde{B}}(x^k)\right)}$$

5.2.3.6 Meta-Consensus

Meta Consensus Net, (see 5.1–5.5), is a particularly suitable and effective new fuzzy function composing sensitivity and precision of each cell of the confusion matrix of each base classifier. The Meta Consensus function was explicitly inspired by Consensus Theory (Tastle et al. 2005; Tastle and Wierman 2007). For each output cell in the base classifiers a weight as calculated on a Meta-Classifier input node. Given k base classifiers in which i and j are subscripts that identify the column (precision) and row (sensitivity) values, (5.6) gives the weight provided by the row calculation.

$$r_{i,j}^k = \frac{a_{i,j}^k}{R_i^k} \cdot \log_2\left(R_i^k - \frac{\left|a_{i,j}^k - R_i^k\right|}{2 \cdot (N-1)}\right); \text{ where } R_i^k = \sum_j^N a_{i,j}^k.$$

(5.6)

A similar weight calculation (5.7) is made based on column values:

$$c_{i,j}^k = \frac{a_{i,j}^k}{C_j^k} \cdot \log_2\left(C_j^k - \frac{|a_{i,j}^k - C_j^k|}{2 \cdot (N-1)}\right); \text{ where } C_j^k = \sum_i^N a_{i,j}^k. \tag{5.7}$$

The data that are missing from these calculations of weight are addressed in an additional weight equation that captures this missing information. Note that from the sum of the rows is subtracted the individual value from the confusion matrix classifier (5.8) to yield the remaining information that is also used to calculate the weight:

$$m_{i,j}^k = \frac{R_i^k - a_{i,j}^k}{R_i^k} \cdot \log_2\left(R_i^k - \frac{a_{i,j}^k}{2 \cdot (N-1)}\right); \tag{5.8}$$

In the same manner that the missing information is calculated for the rows, (5.9) captures the missing information from the column:

$$f_{i,j}^k = \frac{C_j^k - a_{i,j}^k}{C_j^k} \cdot \log_2\left(C_j^k - \frac{a_{i,j}^k}{2 \cdot (N-1)}\right). \tag{5.9}$$

According to Consensus Theory we can assume the following equivalences:

$$Agr(X,\tau) = 1 + \sum_{i=1}^n p_i \log_2\left(1 - \frac{|X_i - \tau|}{2 \cdot d_X}\right) = r_{i,j}^k + c_{i,j}^k;$$

$$Dagr(X,\tau) = -\sum_{i=1}^n p_i \log_2\left(1 - \frac{|X_i - \tau|}{2 \cdot d_X}\right) = m_{i,j}^k + f_{i,j}^k;$$

Consequently the following equation should be able to enhance the consensus of the analyzed confusion matrix:

$$y = Cns(X) - Dnt(X) = \left(r_{i,j}^k + c_{i,j}^k\right) - \left(m_{i,j}^k + f_{i,j}^k\right).$$

We have preferred to express this relationship in logarithmic terms:

$$y^* = \left(\ln(r_{i,j}^k) + \ln(c_{i,j}^k)\right) - \left(\ln(m_{i,j}^k) + \ln(f_{i,j}^k)\right) = -\ln\left(\frac{m_{i,j}^k \cdot f_{i,j}^k}{r_{i,j}^k \cdot c_{i,j}^k}\right).$$

So (5.10) is able to synthesize the Consensus Theory and the Theory of independent judges:

$$w_{i,j}^k = -\ln\left(\frac{m_{i,j}^k \cdot f_{i,j}^k}{r_{i,j}^k \cdot c_{i,j}^k}\right)$$ (5.10)

5.3 Genetic Optimization of Meta-Nets

Meta-Net NNs are completely independent from the classifiers that they have to combine to make a better pattern recognition. Each Meta-Net NN, in fact, builds its weights only analyzing the confusion matrix that the classifiers produce in validation phase. But each confusion matrix represents a compact synthesis of the pattern recognition behavior of a generic classifier. Consequently, Meta-Net NNs do not know analytically how each of its base classifiers performs with each pattern of the validation test.

Considering that, if we put in any Meta-Net pool only one classifier, the Meta-Net algorithm will perform in the Prediction phase (not in the Validation one, because this behavior could be trivial) the same performances of the selected classifier. This is an interesting feature, because it testifies that Meta-Net algorithm is able to infer the general behavior of any classifier, only analyzing a synthesis of its behavior in a specific context (the confusion matrix in the validation phase).

Thus we can say that Meta-Net NNs have a specific intelligent mimetic capability. This fact has important consequences. First of all, we can use Meta-Net NNs to reproduce a generic algorithm, whose mathematic is completely unknown, analyzing only a representative sample of the results it is able to achieve. Second, if we put in a Meta-Net pool a set of classifiers that have shown very different performances in the validation phase, the worst performance of Meta-Net in the prediction phase will be at least equal to the performance of the best classifier in its pool.

A practical way to use this suitable feature is to let each Meta-Met be free to chose its best combination of classifiers in any recognition task. This possibility does not affect the validity of the experimentations because the weights of every Meta-Net are built during the validation phase and they are not changed during the prediction phase. We simply test each Meta-Net in the prediction phase many times, considering all kinds of classifier permutations.

This permutation task can be easy when the number of classifiers is small, but it could be prohibitive when it becomes big. The number of possible permutation, in fact, is determinate by the following equation:

$$NumberOfPermutation = 2^{NumberOfClassifiers}$$

To reduce this CPU time, we have used an evolutionary algorithm called GenD (Buscema 2004; Buscema et al. 2005), that is able to optimize the classifiers pool in

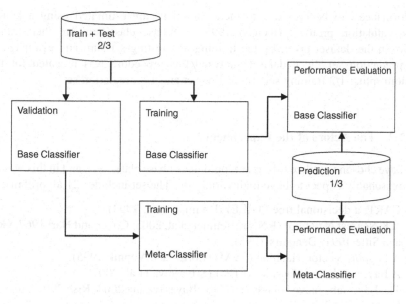

Fig. 5.2 The experimental set up

just a few generations (in this experiment, from 1 to 20 generations at the top, with an average of four generations).

Consequently, each Meta-Net NN, after its weights definition in the validation phase, will use an evolutionary algorithm to select its best combination of classifiers for the prediction phase.

5.3.1 Experimentations

5.3.1.1 Experimental Setup

Figure 5.2 shows the experimental set up carried out to evaluate the results of the Base and of the Meta Classifiers.

Part of the available dataset is used for the training of base classifiers (the training set), part for their performance estimation (the testing set) constituting the informative basis on which the Meta-Classifiers define their strategies of resulting combinations of base classifiers, and a third part (the prediction set) on which the evaluation of the meta-classifiers is conducted. To obtain a correct comparison the base classifiers and the meta-classifiers will be evaluated on the same dataset, the prediction set. The problem with this protocol is that the information, that is to say, the records, available for the training are favorable to meta-classifiers that utilize, through base classifiers, both the training set and the testing set. To correct this unbalance we chose the protocol in Fig. 5.2: the base classifiers' performances, to be provided to meta-classifiers, are estimated using a tenfold cross-validation protocol (Kohavi 1995), the base classifiers are then trained again on the dataset given by the training and testing set sum and evaluated on the prediction set. Each dataset analyzed this procedure was repeated for five random splits: 1/3 training set, 1/3 testing set and 1/3 prediction set.

5.3.1.2   The Actors of the Experiment

We have chosen to utilize different typologies of base classifiers, and in this way we can reasonably expect to have high variability. The set includes 21 algorithms:

1. CART, a decisional tree (TREE) (Breiman et al. 1993),
2. A K Nearest Neighbor (KNN)(Bremner et al. 2005; Cover and Hart 1967; Geva and Sitte 1991; Denoeux 1995),
3. A Support Vector Machine (SVM) (Cortes and Vapnik 1995),
4. A Bayesian linear classifier (LDC) (Srivastava et al. 2007),
5. The Bayesian "naive" classifier (NaiveBayes) (Zang 2004; Rish 2001; John and Langley 1995),
6. A quadratic Bayesian classifier (QDC),
7. An improved Bp algorithm: Delta Bar Delta (DBD) (Jacobs 1988; Patterson 1996),
8. Parzen Classifier (PARZEN) (Parzen 1962; Chapelle 2005),
9. Learning Vector Quantization (LVQ) (Kohonen 1990),
10. A BackPropagation Neural Networks (BP) (Rumelhart et al. 1986; Buscema 1998c),
11. A Sine Network (SN) (Buscema et al. 2006),
12. A Feed Forward Contractive Map (FF_CM) (Buscema and Benzi 2011),
13. Bayes Net (Pearl 1988; Friedman et al. 1997; Holbech and Nielsen 2008; Neal 1996),
14. IBk (Aha et al. 1991; Turney 1993),
15. J48 (LeFevre et al. 2006; Quinlan 1993, 1996),
16. KStar (Cleary and Trigg 1995),
17. Classic Multilayer Perceptron (MPL) (Rumelhart et al. 1986),
18. Multinomial Naïve Bayes (MNB) (Kibriya et al. 2005),
19. Random Forest (Breiman 2001; Livingston 2005),
20. Radial Basis Function (RBF) (Moody and Darken 1989; Powell 1985),
21. Sequential Minimal Optimization (SMO) (Platt 2000;Keerthi et al. 2001).

With respect to Meta-classifiers we propose, for comparison purposes, to compare Meta-Net algorithms with the following 18 algorithms that are among the most used in literature:

1. Wernecke Fusion (Wernecke 1992),
2. Dempster and Shafer Combination (Rogova 1994),
3. Decision Template (Kuncheva 2001; Kuncheva et al. 2001),

4. Majority Vote (MajVote) (Kuncheva 2004; Kittler et al. 1998; Day 1988),
5. Clustering and Selection (Kuncheva 2000),
6. Direct Knn Decision Dependent (DynDdDirectKnn) (Woods et al. 1997),
7. Fuzzy Integral (Cho and Kim 1995),
8. Naïve Bayesan Combiner (BayesComb) (Rokach and Mainon 2001; Stefanowski and Nowaczyk 2006),
9. Weighted Average (Liu 2005),
10. Meta-AdaBoostM1 (Breiman 1998b; Kamath et al. 2001; Mohammed et al. 2006), Freund & Shapire 1997),
11. Meta-Bagging (Breiman 1996),
12. Meta-Dagging (Amasyali and Ersoy 2009),
13. Meta-Decorate (Melville and Mooney 2003),
14. Meta-End (Dong et al. 2005),
15. Meta-LogitBoost (Friedman et al. 1997),
16. Meta-Random Committee (Zorkadis et al. 2005),
17. Meta-Random Sub-Space (Ho 1998),
18. Meta-Rotation Forest (Rodriguez et al. 2006).

For courtesy, we remind also the names of Meta-Net algorithms that we have presented in this research:

1. Meta-Bayes (Buscema 1998e),
2. Meta-Sum
3. Meta-Fuzzy
4. Meta-Exp
5. Meta-Einstein
6. Meta-Consensus (Buscema et al. 2010).

All the algorithms used in this paper were implemented in Matlab (2005), in Neuralware (1998), in WEKA (Hall et al. 2009) and in Semeion Software library (Buscema 1999–2010, 2008–2010).

5.3.1.3   The Datsets

For the experiment we use six datasets derived from the UCI Repository (Asuncion and Newman 2007) and are used in the machine learning area to evaluate the different algorithm performances (Digit, Faults (Buscema et al. 1999), DNA, Letters, Sat-Image and Segment).

5.3.1.4   DIGITS Dataset

From UCI Repository: Semeion Handwritten Digit Dataset.
The dataset used in the problem of recognizing handwritten numeric characters is composed of 1594 digits, handwritten by different subjects in different situations,

and codified in a 256 bit streak corresponding to a 16×16 grid. The objective is to classify each grid into the corresponding digits, 0–9.

5.3.1.5   FAULTS Dataset

From UCI Repository: Steel Plates Faults.

Every dataset record represents a superficial fault of a stainless steel leaf. There are seven different typologies of faults:

1. Pastry;
2. Z_Scratch;
3. K_Scatch;
4. Stains;
5. Dirtiness;
6. Bumps;
7. Other_Faults.

The fault description is constituted by 27 indicators representing the geometric shape of the fault and its contour:

We have 1941 records in total. This dataset was already analyzed in a previous paper (Buscema 1998e).

5.3.1.6   DNA Dataset

From UCI Repository: Molecular Biology (Splice-junction Gene Sequences) Data Set.

Splice junctions are points on a DNA sequence at which "superfluous" DNA is removed during the process of protein creation in higher organisms. The problem posed in this dataset is to recognize, given a sequence of DNA, the boundaries between exons (the parts of the DNA sequence retained after splicing) and introns (the parts of the DNA sequence that are spliced out). This problem consists of two subtasks: recognizing exon/intron boundaries (referred to as EI sites), and recognizing intron/exon boundaries (IE sites). (In the biological community, IE borders are referred to a "acceptors" while EI borders are referred to as "donors".) Details on the dataset can be found at http://archive.ics.uci.edu/ml/machine-learning-databases/molecular-biology/splice-junction-gene-sequences/splice.names. We have used the dataset version with 180 inputs, three outputs and 3,186 data-points.

5.3.1.7 LETTERS Dataset

UCI: Letter Recognition Data Set

The objective is the identification of a great number of boxes containing white and black pixels representing one of the 26 letters of the English alphabet. The characters are extracted by 20 different fonts and distorted in random ways producing 20,000 different characters. Each character has been codified with 16 numeric attributes, scaled on 16 integer values from 0 to 15.

5.3.1.8 SEGMENT Dataset

UCI: Statlog (Image Segmentation) Data Set

The records have been randomly extracted from a database of seven outdoor pictures. These images have been manually sectioned to create a classification for each pixel. Each record represents a 3×3 region. There are seven classes ($1 =$ brick face, $2 =$ sky, $3 =$ foliage, $4 =$ cement, $5 =$ window, $6 =$ path, $7 =$ grass). Every region is characterized by 19 measures on the color image. There are 2,310 records.

5.3.1.9 SAT-IMAGE Dataset

UCI: Statlog (Image Segmentation) Data Set

The database consists of the multi-spectral values of pixels in 3×3 neighborhoods in a satellite image, and the classification is defined by the central pixel, associated with six possible types of soil: red soil, cotton crop, grey soil, damp grey soil, soil with vegetation stubble, and very damp grey soil. The aim is to predict this classification, given the multi-spectral values (four frequencies for each image). The dataset is composed of 36 inputs, 6 outputs and 6,435 data-points.

5.4 The Philosophy of the Experimental Design

5.4.1 The Base Classifiers

We have chosen 21 base classifiers for this experiment. Many of them represent the most popular and most used algorithms for pattern recognition. But they also represent an implicit typology of the Machine Learning world over the last

40 years (Bishop 1995; Ripley 1996; Duda et al. 2001; Witten and Frank 2005; Wang 2010).

In fact, all the algorithms considered in this paper present a suitable feature that is to be learned from data. If we define learning as a process of cognitive manipulation whose target (when the system has learnt) is to reduce to zero the time of the processing itself, then we have to recognize the conceptual specificity of learning: learning is a time process supported by examples and by iteration through which the similarity and the differences among examples are internalized and approximated along the time.

Simultaneously we need to distinguish between these different learning methods:

1. Artificial Neural Networks (ANNs) learning: a bottom-up process to build weight matrices representing the abstract parameters of the data. They are able to compute any kind of non linear function [85–100] (Anderson and Rosenfeld 1988; Arbib 1995; Bishop 1995; Buscema 1998a, b, d; Carpenter and Grossberg 1991; Chauvin and Rumelhart 1995; Hopfield 1988; Kohonen 1995; McClelland and Rumelhart 1988; NeuralWare 1995; Poggio and Girosi 1994; Rumelhart and McClelland 1986; Simpson 1996). Their weights can be linked explicitly to the data features, as the morphemes in natural language, or their weights matrices can represent a distributed and sparse abstraction of the same data, as the phonemes in natural languages.

   (a) Our first type of ANN is the Radial Basis Function (RBF) and Learning Vector Quantization (LVQ). The weights matrices of this ANN are linked to hidden units, each one representing an abstract prototype of the input features;

   (b) The second type of ANN is represented by different types of Back Propagation as the classic Multilayer Perceptron (MLP), an advanced Back Propagation (BP) and the Delta Bar Delta (DBD); in this type we have also included two new ANNs: the Sine Net (SN) and a Feed Forward Contractive Map (FF_Cm). The latter is very important because it shows the possibility of building these kinds of ANNs without utilizing the usual inspiration related to the gradient method. In this case the weight matrices are linked to one or more hidden layers whose units present a distributed coding of the input features. In this way they can be perceived as being more brain like and efficient from the computational point of view. On the other hand, their decoding is very complex and sometime they can appear to be a magic black box. This problem can be actually overcome by using some evolutionary algorithm able to discover the complex pattern through which they coded the input patterns into their hidden units.

2. Function Optimization: in this class we identify all the top-down algorithms whose target is the optimization of a specific cost function. In this case it is necessary to distinguish two subsets of algorithms:

   (a) The Optimization based on Bayesian theory: these algorithms try to optimize some loss function linked strictly to the Bayes theory of probability. The attraction that these algorithms receive from the scholars is due to their mathematic evidence and correctness, and to their speed in processing the

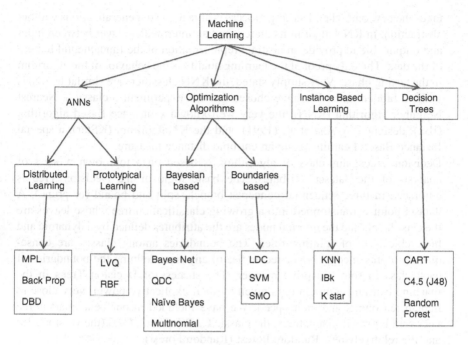

Fig. 5.3 Typology of the base classifiers implemented and tested

data. The time flow and iteration processes in these algorithms are not a theoretical need: they use time as a space to test their hypothesis and in some cases to refine and regularize their weight matrices. We have tested the classic Bayes Network (BayesNet), a Quadratic Bayesian Classifier (QDC), a Parzen Classifier (Parzen) and two types of Naïve Bayes: Simple Naïve Bayes (NaiveBayes) and Multinomial Naïve Bayes (MNB).

(b) The Optimization based on Boundaries: these algorithms try to optimize, linearly and non linearly, the separation boundaries among classes using some type of space transformation. We have chosen for this experimentation the Linear Discriminant Classifier (LDC), the Support Vector Machine (SVM) and the Sequential Minimal Optimization (SMO).

3. Instance Based Learning: this class of algorithms is also known as "lazy" classifiers because they are both very simple and effective at the same time. These algorithms follow the k Nearest Neighbor (KNN) philosophy; the training data generates a tessellation of projection space and each testing datum consequently is assigned automatically to the class of its nearest neighbor. The KNN learning is a bottom-up process based on some kind of distance metric. Usually the chosen metric is weighted to avoid giving the same relevance to all the input attributes. Consequently the quality and the quantity of the training examples increase the tessellation accuracy and then the testing quality. KNN algorithms are similar to a "Pavlov machine" in which learning is generated by direct stimulus – reaction process with a reinforcement of a greedy rule: the nearest

takes the new one. Their learning process is able to make generalizations without abstraction. In KNN algorithms there is not an intermediate layer between input and output able to provide an abstract representation of the fundamental feature of the data. The only proof of the learning quality is the behavior of the algorithm in the testing phase. Very simply stated, the KNN classifiers are similar to ANNs without hidden units. We have chosen for this experiment a classic k Nearest Neighbor algorithm (KNN), the very well known k Instance Based algorithm (IBk), designed by Aha et al. (1991), and the K* algorithm (KStar), a special Instance Based Learning using an entropic distance measure.

4. Decision Trees: this class of algorithms is based on a top-down process of analysis of the dataset attributes. The basic logic of the Decision Tree is embraced in three sequential concepts: branch, split and bound (or prune). A dataset point is transformed into a growing classification tree whose leaves are the class labels and the internal nodes are the attributes defined by a dynamic and hierarchical set of splitting rules. The boundaries among classes are consequently represented by hyper-rectangles of any size, defining the boundaries of each class in some adaptive manner. The success of Decision Trees in the machine learning community is given by their explicative power, very close to human reasoning, and their speed. We have included in our benchmark three very well known decision trees: the classic CART (Tree), C4.5 (the version J48) and the relatively new Random Forest (RandomForest).

Figure 5.3 shows the classification tree of 21 base classifiers implemented in our experiments.

5.4.2   The Meta Classifiers

Meta Classifiers can be defined as a set of adaptive systems able to make improved pattern recognition by combining, in some way, the classification task of other adaptive systems. From an historical point of view, two fundamental styles of combining base classifiers are shown to be effective (Kuncheva 2004; Rokach 2009; Valentini and Masulli 2002; Ho 2001):

(a) Many learning strategies in sampling the training dataset use the same algorithm (Boosting Scheme); Boosting Scheme behaves as an expert using the same lens to watch the world from many points of view. At the end of this overview it decides the true shape of the landscape, according to the coherence of its lens.

(b) Another strategy is that of mixing many different algorithms from the same training dataset, Stacking Scheme (Wolpert 1992; Smyth and Wolpert 1999). The Stacking Scheme uses many different lenses to watch the world from the same point of view, and finally its decision depends on some statistics (theory) derived from its observations.

  Obviously, the machine learning literature has shown many examples of hybrid application of these two extreme schemes, but this distinction can help

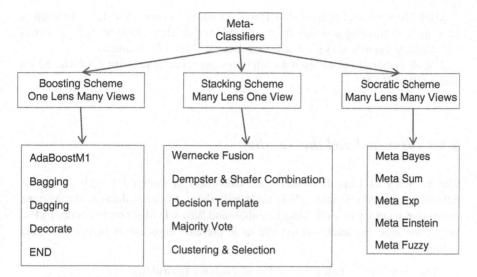

Fig. 5.4 Typology of the 24 Meta-classifiers implemented and tested

the reader to understand how we have chosen to compare the Meta Classifiers described in this paper. Meta Net algorithms could represent a third scheme in building Meta Classifiers:

(c) Many different algorithms, learning locally with many different strategies, could be optimized by one meta-learning algorithm. Meta-Net uses many lens and many points of view, but its meta-learning algorithm is made to be adapted locally to each different base classifier (lens). The Meta-Net target, in fact, is to help each base classifier express its best view and its systematic hallucinations. After this Socratic work, Meta-Net combines, bottom-up, the best observations of all its base classifiers. We have named this approach the Socratic Scheme.

Figure 5.4 shows the 24 Meta Classifiers implemented in this experiment.

5.5   Results

5.5.1   The Numbers of This Experimentation

We have considered 21 base classifiers, 24 Meta-Classifiers and 6 datasets.

Each dataset was processed using a Five Random Split in three subsets: Training & Testing (Tuning set) from one side, and an independent Prediction set from another side. Each classifier and each meta-classifier was tuned in each of the five sessions for any of the six datasets using a K-Fold Cross Validation protocol with $K = 10$. Finally, each one, with tuned parameters, was trained on the Tuning set and tested on the independent Prediction set.

Globally we have implemented 166,320 training sessions and 1,350 prediction sessions[1]: 55 training sessions for each classifier or Meta classifier (21 classifiers and 24 Meta classifiers) in six different datasets (Fig. 5.5a, b and c).

For all elaborations we used an Intel processor, one core, with 2.8 GHz, 32 bit and 4 GB of RAM.

5.5.2 How to Read the Results

The accuracy of Classifiers and Meta-Classifiers is shown for each of the six datasets in one single table. This table will show, for each dataset, the average predictive accuracy of each Meta Classifier and Base Classifier on five independent prediction sets. For each dataset the table presents eight fields in the following columns:

(a) The name or the nick name of the algorithm (Algorithm);
(b) The type of algorithm (Classifier or Meta);

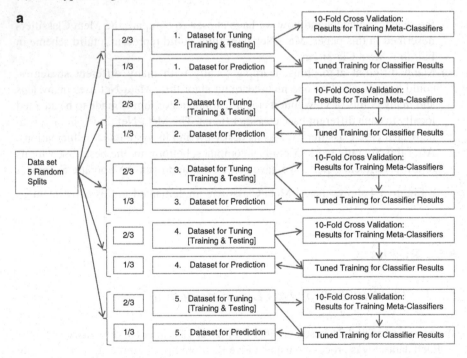

Fig. 5.5 (continued)

---

[1] We gratefully acknowledge Marco Intraligi (Semeion Staff) who helped the authors during the training sessions.

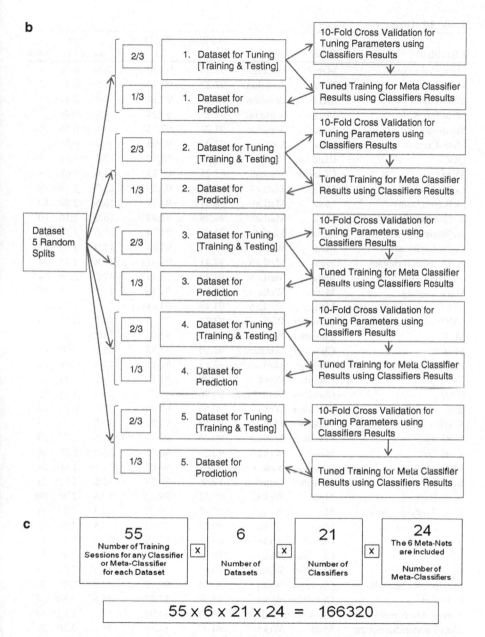

Fig. 5.5 (**a**) Experimental design of any classifier for each dataset; (**b**) experimental design of any Meta classifier for each dataset; (**c**) global number of the training sessions implemented

(c) The software used to implement the algorithm;
(d) The average of the five arithmetic averages of the algorithm in prediction phase (A.Mean);

Table 5.1 Digit dataset – final results

Algorithms	Type	Software	A.Mean (%)	W.Mean (%)	Errors	Var (%)	Rank
Meta-Consensus	Meta	Semeion	97.30	97.30	8.6	0.36	1
Meta-Einstein	Meta	Semeion	97.23	97.24	8.8	0.47	2
Meta-Fuzzy	Meta	Semeion	97.23	97.24	8.8	0.47	2
Meta-Bayes	Meta	Semeion	97.23	97.24	8.8	0.41	2
Meta-Sum	Meta	Semeion	97.04	97.05	9.4	0.36	5
Meta-Expn	Meta	Semeion	97.04	97.05	9.4	0.36	5
QDC	Classifier	MatLab	95.79	95.79	13.4	0.71	7
SVM	Classifier	MatLab	95.73	95.73	13.6	1.08	8
mcMajVote	Meta	MatLab	95.27	95.29	15.0	1.09	9
mcBayesComb	Meta	MatLab	94.81	94.79	16.6	2.49	10
mcDempsterShafer	Meta	MatLab	94.78	94.79	16.6	1.10	10
mcWernecke	Meta	MatLab	94.76	94.79	16.6	1.17	10
Meta-AdaBoostM1	Meta	Weka	94.64	94.66	17.0	1.19	13
mcDecisionTemplate	Meta	MatLab	94.41	94.41	17.8	1.15	14
mcDynDdDirectKnn	Meta	MatLab	94.40	94.41	17.8	2.03	14
Meta-Bagging	Meta	Weka	94.14	94.16	18.6	1.61	16
mcClusteringAndSelection	Meta	MatLab	93.38	93.41	21.0	1.74	17
SMO	Classifier	Weka	93.32	93.35	21.2	1.54	18
MLP	Classifier	Weka	93.00	93.03	22.2	0.91	19
FF_Cm	Classifier	Semeion	92.50	92.53	23.8	2.05	20
Parzen	Classifier	MatLab	92.17	92.22	24.8	1.70	21
Kstar	Classifier	Weka	91.08	91.15	28.2	1.11	22
DBD	Classifier	Neuralware	90.76	90.77	29.4	1.41	23
IBk	Classifier	Weka	90.50	90.59	30.0	0.88	24
KNN	Classifier	MatLab	90.50	90.59	30.0	0.79	24
LDC	Classifier	MatLab	90.26	90.27	31.0	2.37	26
Meta-End	Meta	Weka	90.26	90.27	31.0	1.89	26
LVQ	Classifier	Neuralware	90.07	90.08	31.6	1.86	28
Bp	Classifier	Semeion	89.56	89.58	33.2	1.63	29
Meta-Dagging	Meta	Weka	89.51	89.52	33.4	2.05	30
Meta-RandomCommitte	Meta	Weka	89.15	89.20	34.4	2.93	31
Meta-RotationForest	Meta	Weka	89.05	89.08	34.8	2.50	32
SN	Classifier	Semeion	88.88	88.89	35.4	2.16	33
RBF	Classifier	Weka	88.13	88.13	37.8	2.15	34
RandomForest	Classifier	Weka	86.67	86.75	42.2	3.48	35
Multinomial NaiveBayes	Classifier	Weka	86.16	86.12	44.2	2.49	36
BayesNet	Classifier	Weka	85.39	85.37	46.6	2.80	37
NaiveBayes Simple	Classifier	MatLab	85.01	84.99	47.8	2.82	38
Meta-RandomSubSpace	Meta	Weka	84.77	84.81	48.4	2.36	39
mcFuzzyIntegral	Meta	MatLab	83.36	83.49	52.6	1.66	40
Meta-LogitBoost	Meta	Weka	81.61	81.67	58.4	2.99	41
Meta-Decorate	Meta	Weka	81.19	81.29	59.6	4.68	42
J48	Classifier	Weka	76.60	76.71	74.2	4.02	43
Tree	Classifier	MatLab	73.23	73.32	85.0	4.12	44
mcWeightedAverage	Meta	MatLab	61.09	61.20	123.4	41.63	45

Table 5.2 Faults dataset – final results

Algorithms	Type	Software	A.Mean (%)	W.Mean (%)	Error (%)	Var %	Rank
Meta-Einstein	Meta	Semeion	81.41	80.228	76.8	1.52	1
Meta-Consensus	Meta	Semeion	80.19	80.22	76.8	1.35	1
Meta-Sum	Meta	Semeion	81.26	80.07	77.4	1.42	3
Meta-Fuzzy	Meta	Semeion	81.06	79.91	78.0	1.44	4
Meta-Expn	Meta	Semeion	81.52	79.81	78.4	1.67	5
Meta-Bayes	Meta	Semeion	79.90	79.75	78.6	1.18	6
Meta-AdaBoostM1	Meta	WEKA	80.48	79.65	79.0	2.35	7
Meta-Decorate	Meta	WEKA	79.48	79.14	81.0	1.66	8
Meta-Bagging	Meta	WEKA	78.65	78.47	83.6	1.46	9
Meta-RotationForest	Meta	WEKA	77.63	77.85	86.0	1.46	10
Meta-RandomCommitte	Meta	WEKA	79.52	77.38	87.8	2.42	11
RandomForest	Classifier	WEKA	78.16	77.02	89.2	2.23	12
mcMajVote	Meta	MatLab	80.44	74.76	98.0	2.19	13
Meta-RandomSubSpace	Meta	WEKA	74.62	74.45	99.2	2.80	14
Meta-LogitBoost	Meta	WEKA	74.13	74.40	99.4	0.41	15
SVM	Classifier	MatLab	73.62	74.04	100.8	1.93	16
mcBayesComb	Meta	MatLab	71.95	73.83	101.6	1.86	17
mcDecisionTemplate	Meta	MatLab	79.98	73.73	102	1.82	18
mcDempsterShafer	Meta	MatLab	80.58	73.68	102.2	1.61	19
mcWernecke	Meta	MatLab	78.11	73.37	103.4	1.78	20
TREE	Classifier	MatLab	76.22	73.11	104.4	1.75	21
mcDynDdDirectKnn	Meta	MatLab	77.40	72.59	106.4	1.95	22
FF_Cm	Classifier	Semeion	74.92	72.49	106.8	1.17	23
Kstar	Classifier	WEKA	74.52	71.15	112.0	4.09	24
mcClusteringAndSelection	Meta	MatLab	74.24	70.99	112.6	1.38	25
Parzen	Classifier	MatLab	74.15	70.94	112.8	1.80	26
KNN	Classifier	MatLab	73.62	70.94	112.8	2.18	26
IBk	Classifier	WEKA	73.40	70.94	112.8	2.49	26
mcFuzzyIntegral	Meta	MatLab	77.82	70.74	113.6	0.87	29
SN	Classifier	Semeion	74.16	70.68	113.8	0.79	30
Bp	Classifier	Semeion	74.54	70.53	114.4	1.04	31
MLP	Classifier	WEKA	71.23	70.43	114.8	2.09	32
DBD	Classifier	Neuralware	75.73	70.37	115.0	2.31	33
SMO	Classifier	WEKA	63.09	69.86	117.0	1.97	34
BayesNet	Classifier	WEKA	74.78	69.56	118.2	2.33	35
NaiveBayes Simple	Classifier	MatLab	73.60	68.63	121.8	3.33	36
Meta-End	Meta	WEKA	64.06	68.12	123.8	2.81	37
LVQ	Classifier	Neuralware	74.72	67.28	127.0	2.31	38
Meta-Dagging	Meta	WEKA	55.74	66.67	129.4	2.15	39
RBF	Classifier	WEKA	66.29	65.28	134.8	2.16	40
LDC	Classifier	MatLab	74.25	64.97	136.0	1.81	41
QDC	Classifier	MatLab	77.20	63.37	142.2	1.52	42
J48	Classifier	WEKA	60.33	63.17	143.0	1.64	43
Multinomial NaiveBayes	Classifier	WEKA	71.40	60.44	153.6	1.99	44
mcWeightedAverage	Meta	MatLab	75.75	59.92	155.6	3.69	45

Table 5.3 DNA dataset – final results

Algorithms	Type	Software	A.Mean (%)	W.Mean (%)	Error	Var %	Rank
Meta-Fuzzy	Meta	Semeion	97.64	97.74	14.4	0.56	1
Meta-Consensus	Meta	Semeion	97.55	97.74	14.4	0.45	1
Meta-Einstein	Meta	Semeion	97.59	97.71	14.6	0.59	3
Meta-Expn	Meta	Semeion	97.59	97.71	14.6	0.61	3
Meta-Sum	Meta	Semeion	97.51	97.65	15.0	0.58	5
Meta-Bayes	Meta	Semeion	97.39	97.61	15.2	0.48	6
mcDecisionTemplate	Meta	MatLab	96.69	96.70	21.0	0.61	7
mcWeightedAverage	Meta	MatLab	96.63	96.64	21.4	0.62	8
mcBayesComb	Meta	MatLab	96.42	96.61	21.6	0.53	9
mcDempsterShafer	Meta	MatLab	96.63	96.61	21.6	0.54	9
QDC	Classifier	MatLab	96.51	96.45	22.6	0.62	11
SVM	Classifier	MatLab	95.94	96.23	24.0	0.54	12
mcMajVote	Meta	MatLab	96.17	96.04	25.2	0.74	13
Meta-RotationForest	Meta	Weka	95.10	95.79	26.8	1.35	14
Meta-Bagging	Meta	Weka	94.33	95.39	29.4	0.75	15
mcDynDdDirectKnn	Meta	MatLab	95.20	95.32	29.8	0.46	16
Meta-Decorate	Meta	Weka	94.17	95.04	31.6	0.74	17
MLP	Classifier	WEKA	95.02	95.04	31.6	0.76	17
BayesNet	Classifier	WEKA	94.33	94.85	32.8	0.44	19
mcClusteringAndSelection	Meta	MatLab	94.92	94.82	33.0	0.84	20
Meta-AdaBoostM1	Meta	Weka	94.69	94.79	33.2	1.10	21
NaiveBayes Multinomial	Classifier	WEKA	94.00	94.70	33.8	0.43	22
Meta-End	Meta	Weka	94.25	94.54	34.8	0.80	23
DBD	Classifier	Neuralware	94.34	94.51	35.0	0.82	24
RBF	Classifier	WEKA	94.30	94.22	36.8	0.82	25
Meta-LogitBoost	Meta	Weka	94.23	94.22	36.8	0.68	25
NaiveBayes Simple	Classifier	MatLab	93.55	94.04	38.0	0.87	27
mcFuzzyIntegral	Meta	MatLab	93.95	94.01	38.2	0.75	28
FF_Cm	Classifier	Semeion	93.95	93.97	38.4	0.37	29
Meta-RandomSubSpace	Meta	Weka	93.09	93.97	38.4	0.67	29
Meta-Dagging	Meta	Weka	93.41	93.79	39.6	0.66	31
BP	Classifier	Semeion	93.35	93.69	40.2	0.82	32
LDC	Classifier	MatLab	94.41	93.41	42.0	0.36	33
SN	Classifier	Semeion	92.77	93.19	43.4	0.56	34
SMO	Classifier	WEKA	92.25	93.03	44.4	0.87	35
TREE	Classifier	MatLab	92.21	93.03	44.4	0.96	35
J48	Classifier	WEKA	91.51	92.09	50.4	1.07	37
Meta-RandomCommittee	Meta	Weka	90.30	91.15	56.4	0.33	38
RandomForest	Classifier	WEKA	89.33	90.62	59.8	0.99	39
mcWernecke	Meta	MatLab	88.92	89.83	64.8	1.29	40
KNN	Classifier	MatLab	87.12	88.04	76.2	1.28	41
LVQ	Classifier	Neuralware	78.13	82.70	110.2	0.49	42
Kstar	Classifier	WEKA	78.61	75.24	157.8	0.91	43
IBK	Classifier	WEKA	78.01	73.79	167.0	1.33	44
Parzen	Classifier	MatLab	78.45	73.63	168.0	0.88	45

Table 5.4 Letters dataset – final results

Algorithms	Type	Software	A.Mean (%)	W.Mean (%)	Error	Var %	Rank
Meta-Consensus	Meta	Semeion	98.29	98.31	67.8	0.24	1
Meta-Bayes	Meta	Semeion	98.28	98.29	68.4	0.30	2
Meta'Einstein	Meta	Semeion	98.26	98.28	68.8	0.26	3
Meta-Sum	Meta	Semeion	98.26	98.27	69.2	0.25	4
Meta-Expn	Meta	Semeion	98.25	98.26	69.6	0.25	5
Meta-Fuzzy	Meta	Semeion	98.24	98.25	69.8	0.22	6
SVM	Classifier	MatLab	97.87	97.89	84.6	0.38	7
mcWeightedAverage	Meta	MatLab	97.48	97.49	100.2	0.28	8
mcWernecke	Meta	MatLab	97.33	97.35	106.2	0.27	9
mcClusteringAndSelection	Meta	MatLab	96.98	96.98	120.8	0.49	10
mcDempsterShafer	Meta	MatLab	96.65	96.66	133.4	0.43	11
mcDecisionTemplate	Meta	MatLab	96.62	96.63	134.8	0.27	12
FF_Cm	Classifier	Semeion	96.49	96.50	139.8	0.51	13
mcMajVote	Meta	MatLab	96.45	96.46	141.6	0.40	14
Parzen	Classifier	MatLab	96.20	96.22	151.4	0.29	15
mcDynDdDirectKnn	Meta	MatLab	96.05	96.07	157.2	0.30	16
DBD	Classifier	Neuralware	95.76	95.78	168.8	0.24	17
Ibk	Classifier	WEKA	95.70	95.72	171.2	0.36	18
Kstar	Classifier	WEKA	95.52	95.54	178.6	0.35	19
mcBayesComb	Meta	MatLab	95.35	95.36	185.4	0.30	20
Meta-AdaBoostM1	Meta	WEKA	95.16	95.18	192.8	0.25	21
KNN	Classifier	MatLab	94.86	94.88	205.0	0.43	22
SN	Classifier	Semeion	94.71	94.74	210.6	0.71	23
Meta-RotationForest	Meta	WEKA	94.70	94.74	210.6	0.51	24
Meta-RandomCommitte	Meta	WEKA	94.55	94.58	217.0	0.48	25
Meta-END	Meta	WEKA	94.34	94.37	225.2	0.40	26
BP	Classifier	Semeion	94.19	94.22	231.4	0.56	27
RandomForest	Classifier	WEKA	94.11	94.14	234.4	0.19	28
LVQ	Classifier	Neuralware	94.06	94.08	236.8	0.42	29
Meta-RandomSubSpace	Meta	WEKA	90.81	90.86	365.8	0.97	30
Meta-Decorate	Meta	WEKA	90.41	90.44	382.6	0.46	31
Meta-Bagging	Meta	WEKA	90.37	90.41	383.6	0.97	32
mcFuzzyIntegral	Meta	MatLab	88.65	88.68	452.8	0.67	33
QDC	Classifier	MatLab	88.49	88.54	458.6	0.42	34
Tree	Classifier	MatLab	87.87	87.89	484.4	0.63	35
J48	Classifier	WEKA	87.25	87.28	508.6	0.34	36
SMO	Classifier	WEKA	81.84	81.94	722.2	0.45	37
MLP	Classifier	WEKA	81.32	81.42	743.2	0.82	38
BayesNet	Classifier	WEKA	74.19	74.25	1030.2	0.25	39
RBF	Classifier	WEKA	73.80	73.90	1044.0	0.81	40
Meta-LogitBoost	Meta	WEKA	73.39	73.45	1062.0	0.48	41
Meta-Dagging	Meta	WEKA	73.17	73.31	1067.4	0.65	42
NaiveBayes Simple	Classifier	MatLab	73.22	73.27	1069.4	0.42	43
LDC	Classifier	MatLab	70.07	70.17	1193.4	0.28	44
Multinomial NaiveBayes	Classifier	WEKA	63.90	64.02	1439	0.42	45

Table 5.5 Sat-Image dataset – final results

Algorithms	Type	Software	A.Mean (%)	W.Mean (%)	Error	Var %	Rank
Meta-Einstein	Meta	Semeion	90.76	93.16	88.0	0.71	1
Meta-Fuzzy	Meta	Semeion	90.82	93.16	88.0	0.63	1
Meta-Consensus	Meta	Semeion	90.85	93.15	88.2	0.61	3
Meta-Bayes	Meta	Semeion	90.83	93.15	88.2	0.57	3
Meta-Sum	Meta	Semeion	90.87	93.08	89.0	0.44	5
Meta-Expn	Meta	Semeion	89.84	92.68	94.2	0.55	6
SVM	Classifier	MathLab	90.29	92.35	98.4	0.28	7
Meta-AdaBoostM1	Meta	WEKA	90.22	92.35	98.4	0.68	7
Meta-Bagging	Meta	WEKA	89.00	91.53	109.0	0.63	9
Meta-RotationForest	Meta	WEKA	88.75	91.45	110.0	0.34	10
mcBayesComb	Meta	MatLab	88.47	91.11	114.4	0.66	11
mcMajVote	Meta	MatLab	89.18	91.05	115.2	0.59	12
mcDempsterShafer	Meta	MatLab	89.26	91.02	115.6	0.61	13
Meta-Decorate	Meta	WEKA	88.47	91.02	115.6	1.56	13
mcWernecke	Meta	MatLab	89.02	90.85	117.8	0.72	15
SN	Classifier	Semeion	89.26	90.83	118.0	0.54	16
mcDecisionTemplate	Meta	MatLab	89.47	90.75	119.0	0.72	17
mcClusteringAndSelection	Meta	MatLab	88.82	90.74	119.2	0.73	18
Meta-RandomCommitte	Meta	WEKA	88.39	90.71	119.6	0.80	19
mcWeightedAverage	Meta	MatLab	88.23	90.69	119.8	0.71	20
FF_Cm	Classifier	Semeion	88.92	90.67	120.0	0.44	21
mcDynDdDirectKnn	Meta	MatLab	88.27	90.52	122.0	0.48	22
BP	Classifier	Semeion	88.88	90.49	122.4	0.51	23
KNN	Classifier	MathLab	88.61	90.47	122.6	1.09	24
Meta-End	Meta	WEKA	87.79	90.36	124.0	0.44	25
RandomForest	Classifier	WEKA	87.57	90.18	126.4	0.71	26
Kstar	Classifier	WEKA	88.55	90.08	127.6	0.97	27
IBk	Classifier	WEKA	88.54	89.96	129.2	0.94	28
Parzen	Classifier	MathLab	89.32	89.89	130.2	0.61	29
MLP	Classifier	WEKA	86.54	89.39	136.6	0.79	30
LVQ	Classifier	NeuralWare	85.58	89.21	138.8	0.78	31
Meta-RandomSubSpace	Meta	WEKA	85.66	88.87	143.2	1.02	32
DBD	Classifier	NeuralWare	86.83	87.75	157.6	0.46	33
QDC	Classifier	MathLab	82.45	86.88	168.8	0.44	34
mcFuzzyIntegral	Meta	MatLab	84.54	86.82	169.6	0.63	35
SMO	Classifier	WEKA	82.54	86.60	172.4	0.96	36
J48	Classifier	WEKA	83.17	86.07	179.2	1.03	37
Meta-LogitBoost	Meta	WEKA	82.38	85.82	182.4	1.00	38
TREE	Classifier	MathLab	82.99	85.38	188.2	0.76	39
Meta-Dagging	Meta	WEKA	79.68	84.89	194.4	0.94	40
RBF	Classifier	WEKA	81.43	83.98	206.2	0.89	41
LDC	Classifier	MathLab	81.79	83.87	207.6	0.37	42
NaiveBaye Simple	Classifier	MathLab	80.33	83.65	210.4	0.39	43
BayesNet	Classifier	WEKA	81.06	81.98	232.0	0.61	44
Multinomial NaiveBayes	Classifier	WEKA	78.48	79.46	264.4	0.73	45

Table 5.6 Segment dataset – final results

Algorithms	Type	Software	A.Mean (%)	W.Mean (%)	Error	Var %	Rank
Meta-Bayes	Meta	Semeion	99.09	99.09	4.2	0.39	1
Meta-Consensus	Meta	Semeion	99.09	99.09	4.2	0.39	1
Meta-Expn	Meta	Semeion	99.09	99.09	4.2	0.39	1
Meta-Einstein	Meta	Semeion	99.05	99.05	4.4	0.45	4
Meta-Fuzzy	Meta	Semeion	99.05	99.05	4.4	0.45	4
Meta-Sum	Meta	Semeion	99.05	99.05	4.4	0.45	4
Meta-RandomCommitte	Meta	WEKA	98.01	98.01	9.2	0.49	7
Meta-AdaBoostM1	Meta	WEKA	97.92	97.92	9.6	0.19	8
mcMajVote	Meta	Matlab	97.84	97.84	10.0	0.64	9
mcDecisionTemplate	Meta	Matlab	97.71	97.71	10.6	0.67	10
mcWernecke	Meta	Matlab	97.71	97.71	10.6	0.56	10
mcDempsterShafer	Meta	Matlab	97.66	97.66	10.8	0.69	12
Meta-End	Meta	WEKA	97.66	97.66	10.8	0.60	12
RandomForest	Classifier	WEKA	97.62	97.62	11.0	0.63	14
Meta-Decorate	Meta	WEKA	97.62	97.62	11.0	0.59	14
Meta-RotattionForest	Meta	WEKA	97.57	97.57	11.2	1.08	16
FF_Cm	Classifier	Semeion	97.53	97.53	11.4	0.74	17
mcWeightedAverage	Meta	Matlab	97.40	97.40	12.0	1.09	18
mcBayesComb	Meta	Matlab	97.36	97.36	12.2	0.72	19
SN	Classifier	Semeion	97.36	97.36	12.2	0.42	19
mcDynDdDirectKnn	Meta	Matlab	97.32	97.32	12.4	0.73	21
Kstar	Classifier	WEKA	97.19	97.19	13.0	0.74	22
SVM	Classifier	MatLab	97.14	97.14	13.2	0.99	23
Meta-Bagging	Meta	WEKA	96.98	96.98	13.9	1.07	24
TREE	Classifier	MatLab	96.97	96.97	14.0	0.74	25
Ibk	Classifier	WEKA	96.97	96.97	14.0	0.65	25
Parzen	Classifier	MatLab	96.92	96.92	14.2	0.65	27
J48	Classifier	WEKA	96.92	96.92	14.2	0.80	27
DBD	Classifier	NeuralWare	96.88	96.88	14.4	0.68	29
KNN	Classifier	MatLab	96.80	96.80	14.8	0.69	30
BP	Classifier	Semeion	96.58	96.58	15.8	0.25	31
MLP	Classifier	WEKA	96.58	96.58	15.8	1.17	31
mcFuzzyIntegral	Meta	Matlab	96.45	96.45	16.4	0.56	33
mcClusteringAndSelection	Meta	Matlab	96.23	96.23	17.4	0.90	34
Meta-RandomSubSpace	Meta	WEKA	95.89	95.89	19.0	1.89	35
Meta-LogitBoost	Meta	WEKA	95.89	95.89	19.0	0.53	35
LVQ	Classifier	NeuralWare	95.50	95.50	20.8	0.99	37
SMO	Classifier	WEKA	92.86	92.86	33.0	1.21	38
BayesNet	Classifier	WEKA	91.69	91.69	38.4	1.89	39
LDC	Classifier	MatLab	91.64	91.64	38.6	1.29	40
NaiveBayes Simple	Classifier	MatLab	90.61	90.61	43.4	1.20	41
Meta-Dagging	Meta	WEKA	87.92	87.92	55.8	1.17	42
QD	Classifier	MatLab	87.75	87.75	56.6	1.68	43
RBF	Classifier	WEKA	87.44	87.44	58.0	1.94	44
Multinomial NaiveBayes	Classifier	WEKA	80.30	80.30	91.0	1.00	45

Table 5.7 Results – final ranking

Rank	Algorithms	Type	Software	DataSets						Mean ranking
				Digits	Faults	DNA	Letters	SatImage	Segment	
1	Meta-Consensus	Meta	Semeion	1	1	1	1	3	1	1.33
2	Meta-Einstein	Meta	Semeion	2	1	3	3	1	4	2.33
3	Meta-Fuzzy	Meta	Semeion	2	4	1	6	1	4	3.00
4	Meta-Bayes	Meta	Semeion	2	6	6	2	3	1	3.33
5	Meta-Expn	Meta	Semeion	5	5	3	5	6	1	4.17
6	Meta-Sum	Meta	Semeion	5	3	5	4	5	4	4.33
7	mcMajVote	Meta	MatLab	9	13	13	14	12	9	11.67
8	SVM	Classifier	MatLab	8	16	12	7	7	23	12.17
9	mcDempsterShafer	Meta	MatLab	10	19	9	11	13	12	12.33
10	mcAdaBoostM1	Meta	Weka	13	7	21	21	7	8	12.83
11	mcDecisionTemplate	Meta	MatLab	14	18	7	12	17	10	13.00
12	mcBayesComb	Meta	MatLab	10	17	9	20	11	19	14.33
13	mcWernecke	Meta	MatLab	10	20	40	9	15	10	17.33
14	Meta-Bagging	Meta	Weka	16	9	15	32	9	24	17.50
15	Meta-RotationForest	Meta	Weka	32	10	14	24	10	16	17.67
16	mcDynDdDirectKnn	Meta	MatLab	14	22	16	16	22	21	18.50
17	FF_Cm	Classifier	Semeion	20	23	29	13	21	17	20.50
18	mcClusteringAndSelection	Meta	MatLab	17	25	20	10	18	34	20.67
19	Meta-Decorate	Meta	Weka	42	8	17	31	13	14	20.83
20	Meta-RandomCommitte	Meta	Weka	31	11	38	25	19	7	21.83
21	mcWeightedAverage	Meta	MatLab	45	45	8	8	20	18	24.00

22	Meta-End	Meta	Weka	26	37	23	26	25	12	24.83
23	RandomForest	Classifier	Weka	35	12	39	28	26	14	25.67
24	SN	Classifier	Semeion	33	30	34	23	16	19	25.83
25	Kstar	Classifier	Weka	22	24	43	19	27	22	26.17
26	DBD	Classifier	Neuralware	23	33	24	17	33	29	26.50
27	Parzen	Classifier	MatLab	21	26	45	15	29	27	27.17
28	IBk	Classifier	Weka	24	26	44	18	28	25	27.50
29	KNN	Classifier	MatLab	24	26	41	22	24	30	27.83
30	MLP	Classifier	Weka	19	32	17	38	30	31	27.83
31	QDC	Classifier	MatLab	7	42	11	34	34	43	28.50
32	Bp	Classifier	Semeion	29	31	32	27	23	31	28.83
33	Meta-RandomSubSpace	Meta	Weka	39	14	29	30	32	35	29.83
34	Meta-LogitBoost	Meta	Weka	41	15	25	41	38	35	32.50
35	mcFuzzyIntegral	Meta	MatLab	40	29	28	33	35	33	33.00
36	SMO	Classifier	Weka	18	34	35	37	36	38	33.00
37	Tree	Classifier	MatLab	44	21	35	35	39	25	33.17
38	LVQ	Classifier	Neuralware	28	38	42	29	31	37	34.17
39	BayesNet	Classifier	Weka	37	35	19	39	44	39	35.50
40	J48	Classifier	Weka	43	43	37	36	37	27	37.17
41	Meta-Dagging	Meta	Weka	30	39	31	42	40	42	37.33
42	RBF	Classifier	Weka	34	40	25	40	41	44	37.33
43	LDC	Classifier	MatLab	26	41	33	44	42	40	37.67
44	NaiveBayes Simple	Classifier	MatLab	38	36	27	43	43	41	38.00
45	Multinomial NaiveBayes	Classifier	Weka	35	44	22	45	45	45	39.50

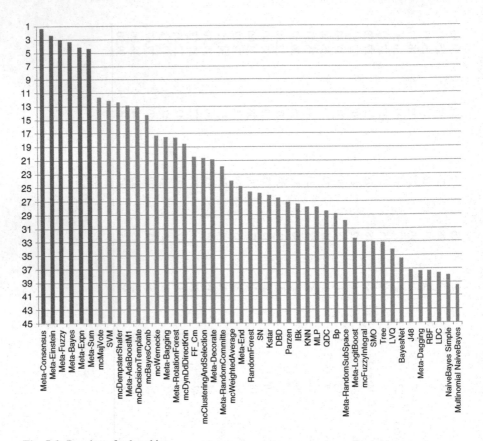

Fig. 5.6 Results – final ranking

(e) The average of the five weighted averages of the algorithm in prediction phase (W.Mean);

(f) The average number of instances misclassified in the five prediction test (Error);

(g) The standard deviation of the five weighted averages of the five predictions test (Var %);

(h) The rank position of the algorithm according to the average of the five weighted averages (Rank).

Because we applied the same procedure to each dataset, we have six tables:

1. Table 5.1: results for Digit dataset;
2. Table 5.2: results for Faults dataset;
3. Table 5.3: results for DNA dataset;
4. Table 5.4: results for Letters dataset;
5. Table 5.5: results for Sat-Image dataset;
6. Table 5.6: results for Segment dataset.

However these results show the obvious power of Meta-Net family, we have summarized the specific performances of each algorithm by the average of its rank position in each dataset. Table 5.7, then, synthesizes the results of this comparison and Fig. 5.6 show the same comparison from a graphical point of view.

5.5.3 Results and Discussion

We have tried to cluster the performances of each classifier and meta-classifier, analyzing its specific misclassifications in the five prediction sets of each dataset (we remind the reader that each algorithm, after the initial training phase, was blindly tested on 30 prediction subsets). Consequently we have built, for each dataset, a square symmetric matrix with null main diagonal where in each cell are counted the number of common patterns that each couple of algorithms misclassified.[2] Obviously we have considered the probability of a couple of classifiers might behave the same, but this depends on the number of output classes and on the number of records of each prediction set. Thus, we have used the following equation to normalize any comparison:

$$S_{i,j} = \frac{2 \cdot (E_i \wedge E_j)}{E_i + E_j} \cdot \frac{K}{R} \cdot \left(1 - \frac{1}{K}\right)$$

where:

E_i and E_j = Number of misclassifications of the i-th or of the j-th classifier;
K = Number of Classes of the Prediction set;
R = Number of Records of the Prediction set;
$S_{i,j}$ = Similarity between the i-th and the j-th classifier in a specific
        Prediction set.

Finally, we have linearly scaled the values of each matrix between 0 and 1 and summed the square matrix over the six datasets.

This final square matrix represents the similarities of misclassifications of each classifier with all the others. Finally, we have calculated the Minimum Spanning Tree (MST) of this matrix in order to generate a weighted graph (tree), clustering the global performances of each algorithm in relation to the others. Figure 5.7a shows the MST of the 45 algorithms behavior.

Some elaboration is necessary:

(a) Meta-Net algorithms are located all together in south-west area of the MST. They are a specific branch of the global tree and they are also strongly

---

[2] This detailed analysis of the shared misclassifications among algorithms was conducted thanks to a suggestion of Dr. Giulia Massini (Semeion researcher).

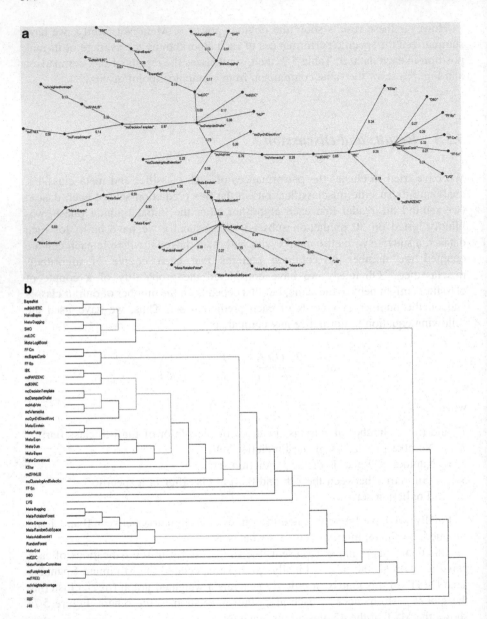

Fig. 5.7 (a) The weighted MST of the similarities in misclassifications of the 45 Algorithms in all six datasets; (b) dendrogram of the similarities in misclassifications of the 45 Algorithm in all the 6 datasets

connected each other. They represent really a new possibility for meta-classifiers;

(b) In the south of the tree most of the Boosting Meta-Classifiers are clustered together. This cluster represents, in a perfect way, the shared philosophy and history of these meta-classifiers. Ada-Boost and Bagging seem to be the prototypes of this group; this position is coherent with the scientific literature.

(c) In the northern area of the tree are clustered the algorithms dedicated to the boundaries optimization. Close to them, in a coherent way, are located the algorithms based on Bayesian optimization.

(d) Instance Base Learning algorithms are grouped together in the eastern side of the tree, with the Parzen classifier bearing a philosophy that is very close to them.

(e) Also in the east, close to Instance Learning algorithms, are clustered the main typology of Artificial Neural Networks. The positions of each of them in the ranking table are very different (see Table 5.7). It is also interesting that the Bayes Combiner is located as its "father node", while all the algorithms based on Bayes theory are located in the opposite side of the tree. We must say that ANNs behave as complex Bayesian machines such that they are not only algorithms based on frequency analysis.

(f) Stacking Meta-Classifiers are spread out into the map, but many of them are located in the central part of the tree: the Majority Vote algorithm, for example, is the center of the graph. Their meaning is quite clear. They, represent an average behavior of the other algorithms. Just an example: Majority Vote, Dempster & Shafer, Decision Template and Wernecke Fusion are located in the center of the tree and are strongly connected each other. That means that they tend to make the same misclassifications and are probably different versions of the same philosophy, based on the voting and /or on the average criterion.

Figure 5.7b shows the dendrogram of the MST of Fig. 5.7a.[3] The dendrogram is useful to understand the dynamics through which the MST was built. In this specific case we have 22 bottom up levels and the first level is the kernel level of MST. This core level is formed by the nine pairs, whose internal similarities are particularly strong:

1. BayesNet and Simple Naïve Bayes;
2. Meta-Dagging and SMO (Meta-Dagging uses SMO as its leading algorithm);
3. FF_Cm and Bayes Combiner (interesting association);
4. IBk and Parzen Classifier;
5. Decision Template and Dempster & Shafer meta classifier;
6. Majority Vote and Wernecke Fusion;
7. Meta-Einstein and Meta-Fuzzy;
8. Meta-Bagging and Meta-Rotation Forest;
9. Fuzzy Integral and Tree (CART).

---

[3] We acknowledge Dr. Giulia Massini (2010–2011) (Semeion researcher) who has generated this dendrogram, using specific software she wrote.

Fig. 5.8 Views of an
unknown system

LEFT SIDE   FRONT

BOTTOM

Fig. 5.9 The 3-dimensional
object with the 2-dimensional
views

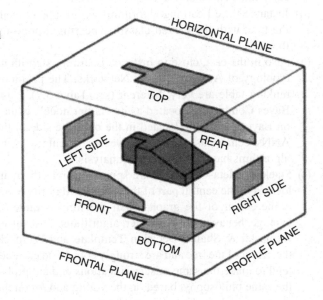

In many cases the nine couples of the core level have a similar position in the
global ranking of the performances (see Table 5.7), but in some cases their
performances are pretty different. In the last case, the more effective algorithm
could substitute for the weaker one.

5.6   Conclusion and Future Works

There is an adage from the field of Systems Science that in order to understand a
system we should all the information necessary to define the structure and behavior
of the system, but no more than what is necessary. Since most all systems have
some kind of flux, possess inputs from unknown or unintended variables and thus
show "noise" in the system, to use more information than that which is necessary to
understand the system is to give the noise an opportunity to play an undeserved role
in the structure and behavior of the system. Further, there are always different views
of a system. One might select a purely mathematical approach to describe a system

while others may produce a model and subject it to an air current to observe the flow. In a similar way each of the Learning Machines described in this paper produce a view of the system from its own specialized perspective, each one correct, but each one only showing a part of the whole system. A simple example will help illustrate the meaning.

Let us examine the different views of an unknown system shown as Fig. 5.8 (from Klir 1985). We have examined the system, using our limited means, to note the shape of the system from the left, the front, and the bottom. Each view is correct, but it is only a view presented by some tool. These are 2-dimensional views of a 3-dimensional figure but each one is decidedly different. Even with something as simple as this set of 2-dimensional views the reader might be challenged to construct the whole system, that is, the 3-dimensional object from which these views, or slices, were taken (see Fig. 5.9). It is only when we assemble the 2-dimensional views that we see the actual or intended structure. These views are orthographic projections when viewed from their respective projection planes, and there are an infinite number of other views depending on the angularity of a plane that passes through the object.

In a similar way each of the Learning Machines and ANNs produce a view that is dependent on the algorithms that analyze the input data. Each Machine Learner is constructed to try to identify the behavior and/or structure of the system, and each does from a particular perspective. It is easy to see above that these Learning Machines result in different answers or views, but each is correct in its own way. There is no evidence to suggest that the results from the other methods might be incorrect, but by using all the information contained in these differing views the Meta-Consensus algorithm utilizes the benefit of having all the views as input to produce the system that is arguably more complete than any of the other individual algorithms. Further, it is difficult to identify which of these individual, single purpose Machine Learners has used more information than necessary to develop its view. It may not be possible to separate the noise from the actual system at the level of the individual network, but it is possible to use the meta-consensus algorithm to assemble the differing views into a single system while disregarding possible noise. The result should be more accurate.

It has been said by some that one must be a *believer* to accept the concept of a neural network or more generally of a learning machine. Perhaps, for the work is being done in the hidden reaches of the computer, even though the resulting output does produce an answer that seems to work quite well. These individual networks demonstrate an ability to solve very complex non-linear problems with relative ease. By combining the output of many stand-alone Learning Machines into one aggregate solution, further strength is given to the output of the meta-consensus neural network.

Thus, we outline the singular properties of the Meta-Net algorithms, especially Meta-Consensus, that we have introduced in this paper. Meta-Net algorithms, in fact, address the pattern recognition problem:

1. The key information that is derived from the algorithms belonging to Meta-Net pool is hidden within details. Machine Learning can be in error because it fixates on the dominant or "correct" relationships (false attributions) and because lapses (missing attributions) can be used to improve the global pattern recognition task of the Meta-Net. Consequently, the marginalities, and then the errors, can frequently be more informative than "correct" behaviors;

2. The real diversity among different Learning Machines is not in their explicit mathematics and topology, but rather, in their performance on real data. In some datasets a Tree can behave in a manner similar to that of an Instance Learning algorithm, or a pool of ANNs can perform in ways that are perceived as similar to that of the Bayesian Combiner. We have learned, for example, that in six different datasets it is the Decision Template and the Dempster-Shafer meta-classifiers that work in practically the same way despite the different theories involved, and that both have similar or inferior performances to that of the simple and easy meta-classifier Majority Vote. Thus we arrive at two inescapable conclusions: first, not everything that shows itself to be different is actually different, and, second (with apologies to William of Ockham), sometime an easy devise works better than a complicated one. To paraphrase, within the *input* we hopefully have the problem, but it is only in the *output* that we have the possibility of discovering truth;

3. Meta-Nets algorithms work bottom up. They follow a majeutic philosophy: they try to allow to any single machine learning of their pool to express completely its point of view about the assigned dataset, and only after they assemble all the different opinions in an open arena where the right solution can arise "spontaneously". The most of the other meta-classifiers follow, instead, two kind of philosophies: an Aristotelian approach (because I know the true, let me see which single machine learning can be useful for that), or a Darwinian approach (The best players survive, the weaker ones die). Sometime "nature" is sweet, and consequently complexity and efficiency diverge.

In conclusion: the future of this meta-solution needs to be illustrated across various disciplines to demonstrate and/or illustrate that these solutions are, overall, a better solution than those done by single purpose networks. This will be a continuing area of research and one with which we are willing to engage in collaborative work with credentialed researchers. Interested researchers are encouraged to contact the corresponding author.

Niels Bohr used to say that an expert is a normal person who made all the possible mistakes only in one field.

References

Aha DW, Kibler D, Albert MK (1991) Instance-based learning algorithms. Mach Learn 6:37–66
Amasyali MF, Ersoy OK (2009) A study of meta learning for regression. ECE technical reports, paper 386. http://docs.lib.purdue.edu/ecetr/386

Anderson JA, Rosenfeld E (eds) (1988) Neurocomputing foundations of research. The MIT Press, Cambridge, MA

Arbib MA (ed) (1995) The handbook of brain theory and neural networks. A Bradford book. The MIT Press, Cambridge, MA

Asuncion A, Newman DJ (2007) UCI Machine Learning Repository [http://www.ics.uci.edu/~mlearn/MLRepository.html]. University of California, School of Information and Computer Science, Irvine

Bishop CM (1995) Neural networks for pattern recognition. Oxford University Press, Oxford

Breiman L et al (1993) Classification and regression trees. Chapman and Hall, Boca Raton

Breiman L (1996) Bagging predictors. Mach Learn 24(2):123–140

Breiman L (1998a) Arcing classifiers. Ann Stat 26(3):801–849

Breiman L (1998b) Arcing classifiers. Ann Stat 26(3):801–849

Breiman L (2001) Random forest. Mach Learn 45(1):5–32

Bremner D, Demaine E, Erickson J, Iacono J, Langerman S, Morin P, Toussaint G (2005) Output-sensitive algorithms for computing nearest-neighbor decision boundaries. Discrete Comput Geometry 33(4):593–604

Buscema M (ed) (1998a) Artificial neural networks and complex social systems – 2. Models, substance use & misuse, vol 33(2). Marcel Dekker, New York

Buscema M (ed) (1998b) Artificial neural networks and complex social systems – 3. Applications, substance use & misuse, vol 33(3). Marcel Dekker, New York

Buscema M (ed) (1998d) Artificial neural networks and complex social systems – 1. Theory, substance use & misuse, vol 33(1). Marcel Dekker, New York

Buscema M, Benzi R (2011) Quakes prediction using a highly non linear system and a minimal dataset. In: Buscema M, Ruggieri M (eds) Advanced networks, algorithms and modeling for earthquake prediction. River Publisher Series in Information Science and Technology, Aalborg

Buscema M (1998e) MetaNet: the theory of independent judges. In: Substance use & misuse, vol 33(2) (Models). Marcel Dekker, Inc., New York, pp 439–461

Buscema M (1999–2010) Supervised ANNs. Semeion software #12, version 16.0

Buscema M (2004) Genetic doping algorithm (GenD): theory and application. Expert Syst 21:2

Buscema M (2008–2010) MetaNets. Semeion Software #44, version 8.0

Buscema M et al (1999) Reti Neurali Artificiali e Sistemi Sociali Complessi. Vol II – Applicazioni. Franco Angeli, Milano, pp 288–291 [Artificial Neural Networks and Complex Systems. Applications]

Buscema M, Grossi E, Intraligi M, Garbagna N (2005) An optimized experimental protocol based on neuro evolutionary algorithms. Application to the classification of dyspeptic patients and to the prediction of the effectiveness of their treatment. Artif Intell Med 34:279–305

Buscema M, Terzi S, Tastle W (2010) A new meta-classifier. In: NAFIPS 2010, 12–14 July, Toronto

Buscema M, Terzi S, Breda M (2006) Using sinusoidal modulated weights improve feed-forward neural networks performances in classification and functional approximation problems. WSEAS Trans Inf Sci Appl 3(5):885–893

Buscema M (1998e) Back propagation neural networks. Subst Use Misuse 33(2):233–270

Carpenter GA, Grossberg S (1991) Pattern recognition by self-organizing neural network. MIT Press, Cambridge, MA

Chapelle O (2005) Active learning for Parzen window classifier. In: Proceedings of the tenth international workshop on artificial intelligence and statistics, pp 49–56. Electronic Paper at http://eprints.pascal-network.org/archive/00000387/02/aistats.pdf

Chauvin Y, Rumelhart DE (eds) (1995) Backpropagation: theory, architectures, and applications. Lawrence Erlbaum Associates, Inc. Publishers, Brodway-Hillsdale

Cho SB, Kim JH (1995) Combining multiple neural networks by fuzzy integral and robust classification. IEEE Trans Syst Man Cybern 25:380–384

Cleary JG, Trigg LE (1995) K*: an instance-based learner using an entropic distance measure. Machine learning inernational conference. Morgan Kaufmann Publishers, San Francisco, pp 108–114

Cortes C, Vapnik V (1995) Support-vector networks. Mach Learn 20(3):273–297

Cover TM, Hart PE (1967) Nearest neighbor pattern classification. IEEE Trans Inf Theory 13 (1):21–27

Day WHE (1988) Consensus methods as tools for data analysis. In: Bock HH (ed) Classification and related methods for data analysis. Elsevier Science Publishers, North Holland, pp 317–324

Denoeux T (1995) A k-nearest neighbor classification rule based on Dempster–Shafer theory. IEEE Trans Syst Man Cybern 25:804–813

Dietterich T (2002) Ensemble learning. In: Arbib MA (ed) The handbook of brain theory and neural networks, 2nd edn. The MIT Press, Cambridge, MA

Dong SL, Frank L, Kramer E (2005) Ensembles of balanced nested dichotomies for multi-class problems. Lecture notes in computer science. NUMB 3721, Springer, Berlin, pp 84–95

Duda RO, Hart PE, Stork DG (2001) Pattern classification. Wiley, New York

Freund Y, Schapire RE (1997) A decision-theoretic generalization of on-line learning and an application to boosting. J Comput Syst Sci 55(1):119–139

Friedman N, Geiger D, Goldszmidt M (1997) Bayesian networks classifiers. Mach Learn 29:131–163

Geva S, Sitte J (1991) Adaptive nearest neighbor pattern classification. IEEE Trans Neural Netw 2 (2):318–322

Hall M, Frank E, Holmes G, Pfahringer B, Reutemann P, Witten IH (2009) The WEKA data mining software: an update. SIGKDD Explorations, 11(1)

Ho TK (1998) The random subspace method for constructing decision forests. IEEE Trans Pattern Anal Mach Intell 20(8):832–844

Ho TK (2001) Data complexity analysis for classifier combination. In: Proceedings of the international workshop on multiple classifier systems. LNCS, vol 2096. Springer, Cambridge, UK, pp 53–67

Holbech S, Nielsen TD (2008) Adapting Bayes network structures to non-stationary domains. Int J Approx Reason 49:379–397

Hopfield JJ (1988) Neural networks and physical systems with emergent collective computational abilities. Proc Natl Acad Sci 79:2554–2558

Jacobs RA (1988) Increased rates of convergence through learning rate adaptation. Neural Netw 1(4):295–307

John GH, Langley P (1995) Estimating continuous distributions in Bayesian classifiers. In: Proceedings of the eleventh conference on uncertainty in artificial intelligence. Morgan Kaufmann Publishers, San Mateo

Kamath C, Cantú-Paz E and Littau D (2001) Approximate splitting for ensembles of trees using histograms. Preprint UCRL-JC-145576, Lawrence Livermore National Laboratory, 1 Oct. US Dept of Energy, Sept 2001

Keerthi SS, Shevade SK, Bhattacharyya C, Murthy KRK (2001) Improvements to Platt's SMO algorithm for SVM classifier design. Neural Comput 13:637–649

Kibriya AM, Frank E, Pfahringer B, Holmes G (2005) Multinomial Naïve Bayes for text categorization revisited. Lecture Notes Comput Sci 3339:235–252

Kittler J, Hatef M, Duin RPW, Matas J (1998) On combining classifiers. IEEE Trans Pattern Anal Mach Intell 20(3):226–239

Klir GJ (1985) Architecture of systems problem solving. Plenum Press, New York

Kohavi R, Provost F (1998) Glossary of terms. Editorial for the special issue on applications of machine learning and the knowledge discovery process, vol 30(2/3)

Kohavi R (1995) A study of cross-validation and bootstrap for accuracy estimation and model selection. In: Proceedings of the fourteenth international joint conference on artificial intelligence. Morgan Kaufmann, San Mateo

Kohonen T (1990) Improved versions of learning vector quantization, 1st edn. International Joint Conference on Neural Networks, San Diego, pp 545–550

Kohonen T (1995) Self-organizing maps. Springer, Berlin

Kuncheva LI (2000) Clustering-and-Selection model for classifier combination. In: Proceedings of the knowledge-based intelligent engineering systems and allied technologies, Brighton, pp 185–188

Kuncheva LI (2001) Switching between selection and fusion in combining classifiers: an experiment. IEEE Trans Syst Man Cybern B 32:146–156

Kuncheva LI (2004) Combining pattern classifiers: methods and algorithms. John Wiley and Sons, Inc., Hoboken, pp 112–125

Kuncheva LI, Bezdek JC, Duin RPW (2001) Decision templates for multiple classifier fusion: an experimental comparison. Pattern Recognit 34(2):299–314

LeFevre K, DeWitt DJ, Ramakrishnan R (2006) Workload Aware Anonymization, KDD'06, Philadelphia, 20–23 Aug

Liu CL (2005) Classifier combination based on confidence transformation. Pattern Recognit 38(1):11–28

Livingston F (2005) Implementing Breiman's random forest algorithm into Weka. ECE591Q machine learning conference papers, 27 Nov

Massini G (2010–2011) MST class. Semeion software #56, ver. 1.0, Rome

Matlab (2005) Version 7

McClelland JL, Rumelhart DE (1988) Explorations in parallel distributed processing. The MIT Press, Cambridge, MA

Melville P, Mooney RJ (2003) Constructing diverse classifier ensembles using artificial training examples. In: Proceedings of the IJCAI-2003, Acapulco, pp 505–510

Mohammed HS, Leander J, Marbach M, Polikar R (2006) Can AdaBoost.M1 Learn Incrementally? A comparison to learn++ under different combination rules. In: Kollias S et al (eds) ICANN 2006, Part I, LNCS 4131, pp 254–263, Springer, Berlin

Moody J, Darken CJ (1989) Fast learning in networks of locally tuned processing units. Neural Comput 1:281–294

Neal RM (1996) Bayesian learning for neural networks. Springer, New York

NeuralWare (1995) Neural computing. NeuralWare Inc., Pittsburgh

NeuralWare (1998) Neuralworks Professional II/Plus, version 5.35

Parzen E (1962) On estimation of a probability density function and mode. Ann Math Stat 33:1065–1076

Patterson D (1996) Artificial neural networks. Prentice Hall, Singapore

Pearl J (1988) Probabilistic reasoning in intelligent systems, representation & reasoning. Morgan Kaufmann Publishers, San Mateo

Platt JC (2000) Probabilistic outputs for support vector machines and comparison to regularized likelihood methods. In: Smola AJ, Bartlett P, Scholkopf B, Schuurmans D (eds) Advances in large margin classifiers. MIT Press, Cambridge, MA

Poggio T, Girosi F (1994) A theory of network of approximation and learning. The MIT Press, Cambridge, MA

Powell MJD (1985) Radial basis function for multi-variable interpolation: a review. IMA conference on algorithms for the approximation of function and data. RMCS, Shrivenham. Also report DAMTP/NA12, Department of Applied Mathematics and Theoretical Physics, University of Cambridge

Quinlan JR (1993) C4.5: programs for machine learning. Morgan Kaufman, San Mateo

Quinlan JR (1996) Improve use of continuous attributes in C4.5. J Artif Intell Res 4:77–90

Ripley BD (1996) Pattern recognition and neural networks. Cambridge University Press, Cambridge, UK

Rish I (2001) An empirical study of the naïve Bayes classifier. In: IBM research report, RC 22230 (W0111-014), New York

Rodriguez JJ, Kuncheva LI, Alonso CJ (2006) Rotation forest: a new classifier ensemble method. IEEE Trans Pattern Anal Mach Intell 28(10):1619–1630

Rogova G (1994) Combining the results of several neural network classifiers. Neural Netw 7:777–781

Rokach L, Mainon O (2001) Theory and applications of attribute decomposition. IEEE international conference on Data Mining, 2001. ICDM 2001, Proceedings IEEE International Conference on 29 Nov-2 Dec 2001

Rokach L (2009) Taxonomy for characterizing ensemble methods in classification tasks: a review and annotated bibliography. Comput Stat Data Anal 53:4046–4072

Rumelhart DE, McClelland JL (eds) (1986) Parallel distributed processing, vol 1. Foundations, explorations in the microstructure of cognition, vol 2. Psychological and biological models. The MIT Press, Cambridge, MA

Rumelhart DE, Hinton GE, Williams RJ (1986a) Learning internal representations by error propagation. In: Rumelhart DE, McClelland JL (eds) Parallel distributed processing, 1st edn, Foundations, Explorations in the Microstructure of Cognition. The MIT Press, Cambridge, MA

Rumelhart DE, Smolensky P, McClelland JL, Hinton GE (1986b) Schemata and sequential thought processes in PDP models. In: McClelland JL, Rumelhart DE, the PDP Group, Parallel distributed processing, vol II. MIT Press, Cambridge, MA, pp 7–57

Simpson P (ed) (1996) Neural networks. Theory, technology, and applications. IEEE Technology Update Series, New York

Smyth P, Wolpert DH (1999) Linearly combining density estimators via stacking. Mach Learn 36:59–83

Srivastava S, Gupta MR, Frigyik BA (2007) Bayesian quadratic discriminant analysis. J Mach Learn Res 8:1277–1305

Stefanowski J, Nowaczyk S (2006) An experimental study of using rule induction algorithm in combiner multiple classifier. Int J Comput Intell Res 2. http://home.agh.edu.pl/~nowaczyk/research/IJCIR.pdf

Tastle WJ, Wierman MJ (2007) Consensus and dissention: a measure of ordinal dispersion. Int J Approx Reason 45(3):531–545

Tastle WJ, Wierman MJ, Dumdum UR (2005) Ranking ordinal scales using the consensus measure. Issues Inf Syst VI(2):96–102

Turney PD (1993) Robust classification with context-sensitive features. In: Proceedings of the sixth international conference on industrial and engineering applications of artificial intelligence and expert systems (IEA/AIE-93), Edinburgh

Valentini G, Masulli F (2002) Ensembles of learning machines. In: Tagliaferri R, Marinaro M (eds) Neural nets, WIRN. Lecture Notes in Computer Science, vol 2486. Springer, Berlin, pp 3–19

Wang PS (ed) (2010) Pattern recognition and machine vision. River Publishers, Aalborg

Wernecke KD (1992) A coupling procedure for the discrimination of mixed data. Biometrics 48(2):497–506

Witten IH, Frank E (2005) Data mining. Morgan Kaufmann, San Francisco

Wolpert DH (1992) Stacked generalization. Neural Netw 5(2):241–259

Woods K, Kegelmeyer WP, Bowyer K (1997) Combination of multiple classifiers using local accuracy estimates. IEEE Trans Pattern Anal Mach Intell 19:405–410

Zang H (2004) The optimality naive Bayes. Am Assoc Artif Intell, www.aaai.org

Zimmermann HJ (1996) Fuzzy set theory and its applications, 3rd edn. Kluwer, Boston/Dordrecht/London

Zorkadis V, Karras DA, Panayotou M (2005) Efficient information theoretic strategies for classifier combination, feature extraction and performance evaluation in improving false positives and false negatives for spam e-mail filtering. Neural Netw 18(5–6), IJCNN 2005, July–Aug, pp 799–807

Chapter 6
Optimal Informational Sorting: The ACS-ULA Approach

Massimo Buscema and Pier Luigi Sacco

6.1 Introduction

Optimal informational sorting is an issue of huge relevance for both theorists and applied scientists. Given an n-dimensional database of characteristics for a list of entities, the correct assignment of each entity to the pertinent group may be far from obvious, provided that the association between carrying certain sets of characteristics and belonging to a specific group is non-linear enough. From the theoretical point of view, problems with this structure can be tackled by a number of different tools, whereas applications practically cover most realms of scientific research, from medical diagnosis, security analysis, and career orientation, to name just a few heterogeneous examples.

In this chapter, we propose a new approach to optimal informational sorting that is based on the combination of two complementary tools: ACS (Activation & Competition System), an auto-associative neural network developed by Massimo Buscema (2009) at Semeion Research Center in Rome, and ULA (Universe Lines Algorithm), a new technique for the generation of "implicit" dynamic datasets that, once coupled to the suitable artificial neural network (ANN) architecture, enhances substantially its discriminating power. This chapter provides a concise but self-contained introduction to these tools, and carries out a validation based on a well known benchmark, widely used in classical approaches to knowledge representation via Parallel Distributed Processing (McClelland 1981, 1995; McClelland and

M. Buscema (✉)
Semeion Research Center of Sciences of Communication, Via Sersale 117, Rome, Italy

Department of Mathematical and Statistical Sciences, CCMB, University of Colorado,
Denver, Colorado, USA
e-mail: m.buscema@semeion.it

P.L. Sacco
IULM University, Via Carlo Bo, 1, 20143 Milano, Italy

W.J. Tastle (ed.), *Data Mining Applications Using Artificial Adaptive Systems*,   183
DOI 10.1007/978-1-4614-4223-3_6, © Springer Science+Business Media New York 2013

Rumelhart 1988), the *West Side Story* dataset where one has to distinguish members' affiliation in two rival gangs, the Jets and the Sharks, on the basis of a certain number of identifying characteristics. This is a demanding benchmark in that characteristics are mixed up in a rather tricky way: Jets tend to be in their 20s, single, and with a Junior High School education, although no one Jet member actually happens to meet all three criteria at the same time, whereas Sharks tend to be older, married, and with a High School education, but again no one Shark happens to meet the three criteria simultaneously. Moreover, all members of both gangs are equally likely to operate as pushers, bookies, or burglars.

The associated sorting problem is particularly hard. We provide a comparative evaluation of different tools for its solution and find that not only the ACS-ULA approach reaches the best results and provides a perfect sorting, but does so by taking as its only input the other methodologies in the benchmark that yield *flawed* results with a *common* bias. Moreover, the ACS-ULA approach allows us to reconstruct in detail not only the affiliation of each member, but also the structure of the association between characteristics and affiliation, including a proper framing of the "deceiving" variables, to identify the "outliers" and to perform an accurate optimal filtering of the dataset. In other words, the input of the approach is not only an accurate sorting, but also an in-depth exploration of the structural properties of the database – that is, very useful information for a number of further tasks, including, for instance, analysis of variance, scenario simulation, vulnerability analysis, and so on. This same procedure may be applied to a vast range of potentially relevant problems in many disciplines.

The structure of the remainder of the chapter is the following. Section 6.2 presents the ACS architecture. Section 6.3 presents the ULA methodology. Section 6.4 introduces the West Side Story database and performs the comparative evaluation. Section 6.5 concludes and offers suggestions for further research.

6.2 Activation and Competition System

ACS is an ANN endowed with an uncommon architecture: Any couple of nodes is not linked by a single value, but by a vector of weights, where each vector component is derived from a specific metric. Such "bio-diversity" of combinations of metrics may provide interesting results when each metric describes different and consistent details of the same dataset. In this situation, ACS forces all the variables to compete among themselves in different respects. The ACS algorithm, therefore, is based on the weights matrices of other algorithms. ACS will use these matrices as a complex set of multiple constraints to update its units in response to any input perturbation. ACS, consequently, works as a dynamic non linear associative memory. Whenever any input is set on, ACS will activate all its units in a dynamic, competitive and cooperative process at the same time. This process will terminate when the evolutionary negotiation among all the units finds its natural attractor. The ACS ANN is, thus, a complex kind of C.A.M. system (Content Addressable Memory). Compared to the classic associative memories (McClelland and

Rumelhart 1988; Hinton and Anderson 1981; Grossberg 1980), ACS presents the following new features:

- The ACS algorithm works using simultaneously many weights matrices coming from different algorithms and/or ANNs;
- The ACS algorithm recall is not a one-shot reaction but an evolutionary process where all its units negotiate their reciprocal value.

To compute the weights matrices for the ACS algorithm, one can follow different approaches: Applying straightforward formulas for association among variables, or making use of more complex algorithms such as specific ANN architectures like Self-Organizing Maps (SOMs) (Kohonen 1995) and Auto-Contractive Maps (AutoCM) (Buscema 2007a, b), as well as any kind of mix of the above. In this chapter, however, for reasons that will become clear later, we will only make use of very simple formulas for association.

6.3 Measures of Association

The matrix of associations of M variables from a dataset with N patterns can easily be constructed by computing the linear associations between any couple of the M variables:

$$W_{i,j}^{[L]} = \frac{\sum\limits_{k=1}^{N} (x_{i,k} - \bar{x}_i) \cdot (x_{j,k} - \bar{x}_j)}{\sqrt{\sum\limits_{k=1}^{N} (x_{i,k} - \bar{x}_i)^2 \cdot \sum\limits_{k=1}^{N} (x_{j,k} - \bar{x}_j)^2}}; \tag{6.1}$$

$$-1 \le W_{i,j}^{[L]} \le 1; \; i,j \in [1,2,...,M]$$

The associations matrix, $W_{i,j}^{[L]}$, is a square matrix in which the main diagonal entries are zero. The matrix $W_{i,j}^{[L]}$ has, however, some limitations. It considers only linear relationships among variables and it is not sensitive to the frequency and to the distribution of the variables across the dataset. To compensate for these limitations, we compute another association matrix, $W_{i,j}^{[P]}$, based on the distribution probability of co-occurrence of any couple of the M variables:

$$W_{i,j}^{[P]} = -\ln \frac{\frac{1}{N^2} \cdot \sum\limits_{k=1}^{N} x_{i,k} \cdot (1 - x_{j,k}) \cdot \sum\limits_{k=1}^{N} (1 - x_{i,k}) \cdot x_{j,k}}{\frac{1}{N^2} \cdot \sum\limits_{k=1}^{N} x_{i,k} \cdot x_{j,k} \cdot \sum\limits_{k=1}^{N} (1 - x_{i,k}) \cdot (1 - x_{j,k})} \tag{6.2}$$

$$-\infty \le W_{i,j}^{[P]} \le +\infty; \; x \in [0,1]; \; i,j \in [1,2,...,M]$$

If we scale linearly this new matrix, $W_{i,j}^{[P]}$, in the same interval as for the linear matrix, $W_{i,j}^{[L]}$, we get two comparable hyper-surfaces into the same metric space.

6.4   The ACS Algorithm

ACS is a nonlinear associator whose cost function is based on the minimization of the energy among units whenever the system is activated by an external input. We call it Activation & Competition System (ACS for short) and it is defined as follows:

M = Number of Variables-Units;

Q = Number of weights matrices;

$i,j \in M$;

$k \in Q$;

$W_{i,j}^k$ = value of connection between the i-th and the j-th units of the k-th matrix;

Ecc_i = global excitation to the i-th unit coming from the other units;

Ini_i = global inhibition to the i-th unit coming from the other units;

E_i = final global excitation to the i-th unit;

I_i = final global inhibition to the i-th unit;

$[n]$ = cycle of the iteration;

$u_i^{[n]}$ = state of the i-th unit at cycle n;

$H^{[n]}$ = amount of units updating at cycle n;

Net_i = Net Input of the i-th unit;

δ_i = delta update of the i-th unit;

$Input_i$ = value of the i-th external input: $-1 \le Input_i \le +1$;

$N_{k,i}^{[E]}$ = number of positive weights of the k-th matrix to the i-th unit;

$N_{k,i}^{[I]}$ = number of negative weights of the k-th matrix to the i-th unit;

Max = Maximum of activation: $Max = 1.0$;

Min = Minimum of activaction: $Min = -1.0$;

$Rest$ = rest value: $Rest = -0.1$;

$Decay_i^{[n]}$ = Decay of activaction the i-th unit at cycle n : $Decay_i^{[n=0]} = 0.1$;

α = scalar for the E_i and I_i net input to each unit;

β = scalar for the external input;

ε = a small positive quantity close to zero.

$$Ecc_i = \alpha \cdot \sum_k \frac{\displaystyle\sum_i^M u_i^{[n]} \cdot W_{i,j}^k}{N_{k,i}^{[E]}} \quad W_{i,j}^k > 0;$$

$$Ini_i = \alpha \cdot \sum_k \frac{\displaystyle\sum_i^M u_i^{[n]} \cdot W_{i,j}^k}{N_{k,i}^{[I]}} \quad W_{i,j}^k < 0;$$

$$E_i = Ecc_i + \beta \cdot Input_i; \quad Input_i > 0;$$

$$I_i = Ini_i + \beta \cdot Input_i; \quad Input_i < 0;$$

$$Net_i = \left(Max - u_i^{[n]}\right) \cdot E_i + \left(u_i^{[n]} - Min\right) \cdot I_i - Dec_i \cdot \left(u_i^{[n]} - Rest\right);$$

$$\delta_i = Net_i \cdot (1.0 - u_i^2);$$

$$H^{[n]} = \sum_i^M \delta_i^2;$$

$$u_i^{[n+1]} = u_i^{[n]} + \delta_i;$$

$$Dec_i^{[n+1]} = Dec_i^{[n]} \cdot e^{-u_i^2}; \tag{6.3}$$

Notice how the decay function of ACS is itself subject to learning: The more pronounced the activation or inhibition of units (i.e., the more u_i gets closer to $+1$ or -1), the quicker decay approaches zero.

$H^{[n]}$ is the cost function of ACS to be minimized. Consequently, when $H^{[n]} < \varepsilon$, the algorithm terminates. More specifically:

$$Max \cdot E_i - u_i \cdot E_i + u_i \cdot I_i - Min \cdot I_i - Dec_i \cdot u_i + Rest \cdot Dec_i = 0$$

$$Max \cdot E_i - u_i \cdot E_i + u_i \cdot I_i - Min \cdot I_i - Dec_i \cdot u_i + Rest \cdot Dec_i = 0$$

$$(-E_i + I_i - Dec_i) \cdot u_i + Max \cdot E_i - Min \cdot I_i + Rest \cdot Dec_i = 0$$

$$u_i = \frac{Max \cdot E_i - Min \cdot I_i + Rest \cdot Dec_i}{E_i - I_i + Dec_i} \tag{6.4}$$

When $Max = 1; Min = -1; Rest = 0.1$, then:

$$u_i = \frac{Ecc_i + I_i - 0.1 \cdot Dec_i}{Ecc_i - I_i + Dec_i} \tag{6.5}$$

We have already said that the ACS ANN is partially inspired by a previous ANN presented by Grossberg (1976, 1978, 1980) but their differences are so marked that we need to present ACS as a new ANN:

- ACS works using simultaneously many weights matrices coming from different algorithms, while Grossberg' IAC uses only one weight matrix;
- ACS weights matrices represent different mappings of the same dataset and all the units (variables) are processed in the same way, while Grossberg's IAC works only when the dataset presents only a specific kind of architecture;
- The ACS algorithm can use any combination of weights matrices coming from any kind of algorithm. The only constraint is that all the values of every weights matrix have to be linearly scaled into the same range (typically between -1 and $+1$), while Grossberg's IAC can work only with static excitations and inhibitions.
- Each ACS unit tries to learn its specific value of decay, during its interaction with the other units, while Grossberg's IAC works with a static decay parameter for all the variables;
- The ACS architecture is a circuit with symmetric weights (vectors of symmetric weights), which can manage a dataset with any kind of variables (Boolean, categorical, continuous, etc.), while Grossberg's IAC can work only with specific types of variables.

The ACS System is implemented by specific research software patented by Semeion Research Center (Buscema 2009).

6.5 Universe Lines

ACS, using the weight matrices provided by different algorithms, transforms a static dataset into a dynamical system: Every external activation of one of the variables of the original dataset generates a process. Each step of this process is a state of a dynamical system. At each state, each variable takes a specific value, generated by the previous evaluation of that variable with all the others. The process will terminate when the system reaches its natural attractor. At this point, all the states of the process represent the trajectory drawn by ACS with reference to the first external input. Thus, for each external recall of the dataset, we get a new dataset.

If the original dataset is made of N variables, after the training phase (generation of weight matrices), we can make N independent recalls, one for any of the N variables. ACS will generate N different datasets, each one representing the path (trajectory) of a specific recall into an $N-1$ dimensional space. Of course, each trajectory could have a different number of states. But it is easy to standardize the number of states for the trajectories making an homogenous sampling from them.

As the process ends, we have transformed a static dataset of N variables into N dynamical datasets (systems), each of which represents a universe line for each variable of the original dataset. The advantages of this transformation are many:

1. Each universe line represents how any specific variable modifies itself and the other variables during a sequential process of negotiation;
2. Each universe line is a trajectory, a sequential machine, whose states cause the competition and the cooperation processes going on among all of the variables to emerge spontaneously;
3. Once this happens, we can analyze the relationship between any couple of variables using two (N−1)-dimensional vectors, and not only one scalar value.

Cropping all of the N universe lines together, we get a new view of the original dataset: The universe lines suggest us how the relationship among variables is going to change. In mathematical terms:

$$L_{i_{NxM_i}}(I_i, \mathbf{x}_i) = \overline{\mathbf{x_i}}; \tag{6.6}$$

$$\mathbf{x}_i^{n+1} = ACS_i(I_i = 1, \mathbf{x}_i^n, \mathbf{w}). \tag{6.7}$$

Legenda:

$N =$ Number of dataset variables;

$\mathbf{M}_i =$ Number of ACS cycles for the ith variable;

$I_i =$ External Input of the ith variable;

$\mathbf{x}_i^n =$ ith Vector of the state of all of the variables at each cycle n;

$\mathbf{w} =$ Constant matrices of weights among the variables;

$ACS_i =$ function of the ith dynamical system;

$\overline{\mathbf{x_i}} =$ Final sequential matrix for the ith variable;

$L_{i_{NxM}}(I_i, \mathbf{x}_i) =$ Universe line of the ith variable in relation to the N − 1 other variables in M(i) cycles

The flow chart of ULA is synthetically depicted in Fig. 6.1 below.

6.6   The West Side Story Dataset

The best way to illustrate the practical and the conceptual advantages of the Universe Line Algorithm (ULA for short) is to work with a test problem that is well studied in order to provide a meaningful and easily comparable benchmark. For this purpose, we have chosen to work with the so called Gang dataset inspired by the well known *West Side Story* characters (Tables 6.1 and 6.2).

We can sum up the basic statistics of the dataset as in Table 6.3 below:

Accordingly, in Tables 6.4 and 6.5 we report the Carnot maps of the frequencies of the whole range of combinations of the variables in their relationships to the characteristics of the two Gangs.

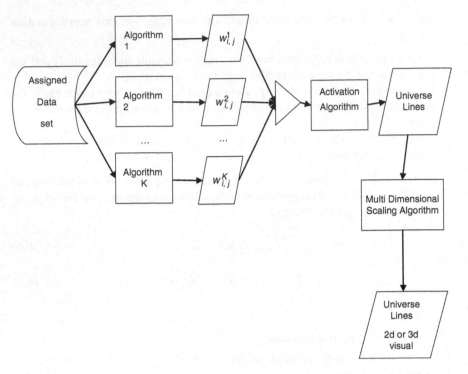

Fig. 6.1 Processing work flow: from the assigned dataset to the visualization of the Universe Lines

6.7 ULA Analysis

As a first step, we will apply a linear correlation analysis to every couple of variables of the dataset. Then, we will use the same linear correlation matrix weights to calculate, through the ACS algorithm, the Universe Lines matrices for every variable. We will uniform the cardinality of the cycles of any new UL matrix, and then merge all the ULs in a new dataset (ULs Dataset). Finally, we will again calculate the linear correlation of this new dataset, so as to compare the first linear correlation matrix with the second. As a first stage, the ACS methodology prescribes that we have to run alternative classification algorithms on the dataset so as to generate the weight matrices that are needed to feed the ACS itself. To this purpose, we have considered two different methodologies: Linear Correlation (LC) and Prior Probability (PP) analysis. LC analysis is based on the first measure of association presented in (6.1) of Sect. 6.1. PP analysis is based on the enhanced version of the "simple" measure of association, as presented in (6.2) of Sect. 6.1.

For each approach, we thus get a weights matrix, mapping in an idiosyncratic way the relationships existing between the variables in the dataset. The weight matrices for the two methodologies are reported in Tables 6.6 and 6.7.

Table 6.1 The West Side Story dataset

The dataset and its basic statistics

Name	Gang	Age	Education	Status	Profession
ART	Jets	40	Junior school	Single	Pusher
AL	Jets	30	Junior school	Married	Burglar
SAM	Jets	20	College	Single	Bookie
CLYDE	Jets	40	Junior school	Single	Bookie
MIKE	Jets	30	Junior school	Single	Bookie
JIM	Jets	20	Junior school	Divorced	Burglar
GREG	Jets	20	High school	Married	Pusher
JOHN	Jets	20	Junior school	Married	Burglar
DOUG	Jets	30	High school	Single	Bookie
LANCE	Jets	20	Junior school	Married	Burglar
GEORGE	Jets	20	Junior school	Divorced	Burglar
PETE	Jets	20	High school	Single	Bookie
FRED	Jets	20	High school	Single	Pusher
GENE	Jets	20	College	Single	Pusher
RALPH	Jets	30	Junior school	Single	Pusher
PHIL	Sharks	30	College	Married	Pusher
IKE	Sharks	30	Junior school	Single	Bookie
NICK	Sharks	30	High school	Single	Pusher
DON	Sharks	30	College	Married	Burglar
NED	Sharks	30	College	Married	Bookie
KARL	Sharks	40	High school	Married	Bookie
KEN	Sharks	20	High school	Single	Burglar
EARL	Sharks	40	High school	Married	Burglar
RICK	Sharks	30	High school	Divorced	Burglar
OL	Sharks	30	College	Married	Pusher
NEAL	Sharks	30	High school	Single	Bookie
DAVE	Sharks	30	High school	Divorced	Pusher

Alternatively, we run the ULA algorithm, taking as inputs for the ACS the output of the two "simple" methodologies, LC and PP. Thus, the ACS-ULA methodology plainly builds on the inferences that can be drawn through straightforward measures of association. The weight matrix for the ACS-ULA methodology is presented in Table 6.8; notice how, unlike the previous ones, the non-zero entries of this weight matrix are entirely made of either full activations (+1) or inhibitions (−1).

One could expect that the accuracy of the latter critically depends on that of the former, and that ACS-ULA performs by yielding relatively marginal, quantitative improvements in terms of accuracy with respect to the simple algorithms. In fact (and this was in fact clear by simply looking at the structure of the weight matrix), this intuition proves to be wrong: ACS-ULA produces a major, qualitative improvement on the inferences of the simple algorithms, even if they turn out to be flawed by a *common* bias. In other words, ACS-ULA improves upon the results of LC and

Table 6.2 The West Side Story dataset, binary format: 14 variables × 27 records

	Jet	Sharks	20s	30s	40s	JH	COL	HS	Single	Married	Divorced	Pusher	Bookie	Burglar
ART	1	0	0	0	1	1	0	0	1	0	0	1	0	0
AL	1	0	0	1	0	1	0	0	1	1	0	0	0	1
SAM	1	0	1	0	0	0	1	0	1	0	0	0	1	0
CLYDE	1	0	0	0	1	1	0	0	1	0	0	0	1	0
MIKE	1	0	0	1	0	1	0	0	1	0	1	0	1	0
JIM	1	0	1	0	0	0	0	1	0	1	0	1	0	1
GREG	1	0	1	0	0	1	0	0	0	0	0	1	0	0
JOHN	1	0	0	1	0	0	0	1	1	1	0	0	0	1
DOUG	1	0	0	1	0	1	0	0	0	1	0	0	1	0
LANCE	1	0	1	0	0	0	0	1	1	0	1	0	0	1
GEORGE	1	0	1	0	0	0	0	1	0	0	0	0	1	1
PETE	1	0	1	0	0	1	0	0	1	0	0	1	0	0
FRED	1	0	1	0	0	0	1	0	1	0	0	1	0	0
GENE	1	0	0	1	0	1	0	0	1	1	0	1	0	0
RALPH	1	0	0	1	0	0	1	0	0	0	0	0	0	0
PHIL	0	1	0	1	0	1	0	0	1	1	0	1	1	0
IKE	0	1	0	1	0	0	1	0	1	0	0	0	0	0
NICK	0	1	0	1	0	0	0	1	0	1	0	1	1	0
DON	0	1	0	1	0	0	1	0	0	1	0	0	0	1
NED	0	1	0	0	1	0	0	1	0	0	1	0	1	0
KARL	0	1	1	0	0	1	0	1	1	1	0	0	1	0
KEN	0	1	0	0	1	0	0	1	0	0	0	0	0	1
EARL	0	1	0	0	0	0	0	1	0	1	0	0	0	1
RICK	0	1	0	1	0	0	1	0	0	0	1	0	0	1
OL	0	1	0	1	0	0	0	1	0	1	0	1	0	0
NEAL	0	1	0	1	0	0	0	1	0	0	0	0	1	0
DAVE	0	1	0	1	0	0	0	1	0	0	1	1	0	0

Table 6.3 The West Side Story dataset, basic statistics	Freq	Jets	Sharks	Frequence	Jets (%)	Sharks (%)
	20s	9	1	20s	60.00	8.33
	30s	4	9	30s	26.67	75.00
	40s	2	2	40s	13.33	16.67
	JH	9	1	JH	60.00	8.33
	HS	4	7	HS	26.67	58.33
	COL	2	4	COL	13.33	33.33
	Single	9	4	Single	60.00	33.33
	Married	4	6	Married	26.67	50.00
	Divorced	2	2	Divorced	13.33	16.67
	Pusher	5	4	Pusher	33.33	33.33
	Bookie	5	4	Bookie	33.33	33.33
	Burglar	5	4	Burglar	33.33	33.33

PP even in cases in which they commit the *same* mistakes. This is the consequence of the dynamic unfolding of information carried out by ULA through the sequential generation of virtual datasets.

To make a precise evaluation of the amount of information generated by the dynamic unfolding carried out by ULA, we will calculate again the linear correlation between the linear correlation of each variable with respect to the others in the original input matrix, and the linear correlation of the same variable with respect to the others in the final matrix. Eventually, as N is the number of variables of the dataset ($N = 14$), we will generate N *meta correlations* among the variables before and after the application of the UL Algorithm.

If the meta-correlation indices turn out to be close to 1 (or to -1), that means the ULA has not extracted new information from the original dataset: An almost perfect correlation (or anti-correlation) reveals that the new weight matrix is a mere perturbation of the original one. Whereas, if such indices are significantly away from (plus or minus) unity, that means that ACS-ULA has generated a substantial amount of new information with respect to the original dataset. In the latter case, we need to understand whether such change improves our knowledge about the Gang dataset or is nothing but noise coding. Table 6.9 presents the linear correlation matrices before and after the application of ULA (with autocorrelation set to 0):

The result of the comparison is clear: The two matrices present a meaningful difference. Their meta-correlation is relatively low (r = 0.5858). They are certainly linked, but weakly enough to be taken as substantially different. Moreover, the two matrices are different in different ways, according to the specific variable we focus upon:

- Some Meta correlations are quite high (Jets, Sharks, JH): This means that, for these variables, ULA does not generate a substantial amount of extra information;
- Some Meta correlations are low (20s, 30s, College, Bookie), keeping in mind that the source dataset is the same and the ULA is a deterministic process; this could mean that ULA has extracted new information from the original dataset;

Table 6.4 Carnot map for Jets

Jets	20 age	20 age	20 age	30 age	30 age	30 age	40 age	40 age	40 age	
	Junior school	High school	College	Junior school	High school	College	Junior school	High school	College	
Single	0	1	1	1	0	0	1	0	0	Pusher
Single	0	0	0	0	0	0	0	0	0	Burglar
Single	0	1	1	1	1	0	1	0	0	Bookie
Married	0	1	0	0	0	0	0	0	0	Pusher
Married	2	0	0	1	0	0	0	0	0	Burglar
Married	0	0	0	0	0	0	0	0	0	Bookie
Divorced	0	0	0	0	0	0	0	0	0	Pusher
Divorced	2	0	0	0	0	0	0	0	0	Burglar
Divorced	0	0	0	0	0	0	0	0	0	Bookie

Table 6.5 Carnot map for Sharks

Sharks	20 age	20 age	20 age	30 age	30 age	30 age	40 age	40 age	40 age	
	Junior school	High school	College	Junior school	High school	College	Junior school	High school	College	
Single	0	0	0	0	1	0	0	0	0	Pusher
Single	0	1	0	0	0	0	0	0	0	Burglar
Single	0	0	0	1	1	0	0	0	0	Bookie
Married	0	0	0	0	0	2	0	0	0	Pusher
Married	0	0	0	0	0	1	0	1	0	Burglar
Married	0	0	0	0	1	1	0	1	0	Bookie
Divorced	0	0	0	0	1	0	0	0	0	Pusher
Divorced	0	0	0	0	1	0	0	0	0	Burglar
Divorced	0	0	0	0	0	0	0	0	0	Bookie

Table 6.6 The LC weights matrix

[LC]	ART	AL	SAM	CLYDE	MIKE	JIM	GREG	JOHN	DOUG	LANCE	GEORGE	PETE	FRED	GENE	RALPH	PHIL	IKE	NICK	DON	NED	KARL	KEN	EARL	RICK	OL	NEAL	DAVE
ART	0.00	0.07	0.07	0.69	0.38	0.07	0.07	0.07	0.07	0.07	0.07	0.07	0.38	0.38	0.69	-0.24	0.07	0.07	-0.56	-0.56	-0.24	-0.24	-0.24	-0.56	-0.24	-0.24	-0.24
AL	0.07	0.00	-0.24	0.07	0.38	0.38	0.07	0.69	0.07	0.69	0.69	-0.24	-0.24	-0.24	0.38	0.07	0.07	-0.24	0.38	0.07	-0.24	-0.24	0.07	0.07	0.07	-0.24	-0.24
SAM	0.07	-0.24	0.00	0.38	0.38	0.07	0.07	0.07	0.38	0.07	0.38	0.69	0.38	0.69	0.07	-0.24	0.07	-0.24	-0.24	0.07	-0.24	0.07	-0.56	-0.56	-0.24	0.00	-0.56
CLYDE	0.69	0.07	0.38	0.00	0.69	0.07	-0.24	0.07	0.38	0.07	0.07	0.38	0.07	0.07	0.38	-0.56	0.38	-0.24	-0.56	-0.24	0.07	-0.24	-0.24	-0.56	-0.56	0.07	-0.56
MIKE	0.38	0.38	0.38	0.69	0.00	0.07	-0.24	0.69	0.69	0.07	0.07	0.38	0.69	-0.24	0.69	-0.24	0.69	0.07	-0.24	0.07	-0.24	-0.24	-0.56	-0.24	-0.24	0.38	-0.24
JIM	0.07	0.38	0.07	0.07	0.07	0.00	0.07	0.69	-0.24	0.69	0.07	0.07	0.07	0.69	0.07	-0.56	-0.24	-0.56	-0.24	-0.56	-0.24	0.07	-0.24	0.07	-0.56	-0.56	-0.24
GREG	0.07	0.07	0.07	-0.24	-0.24	0.07	0.00	0.38	0.07	0.38	0.07	0.38	0.69	0.38	0.07	0.07	-0.56	0.07	-0.24	-0.24	0.07	0.07	0.07	0.38	0.38	-0.24	0.07
JOHN	0.07	0.69	0.07	0.07	0.69	0.69	0.38	0.00	0.07	1.00	-0.24	0.07	0.07	0.07	0.07	-0.24	-0.24	-0.56	0.07	-0.24	0.07	0.07	0.07	-0.24	-0.24	-0.56	-0.56
DOUG	0.07	0.07	0.38	0.38	0.69	-0.24	0.07	0.07	0.00	-0.24	0.00	-0.24	0.38	0.07	0.38	-0.24	0.38	0.38	-0.24	0.07	0.07	-0.24	0.07	-0.24	-0.24	0.69	0.07
LANCE	0.07	0.69	0.07	0.07	0.07	0.69	0.38	1.00	-0.24	0.00	-0.24	0.69	0.07	0.07	0.07	-0.24	-0.24	-0.56	0.07	-0.24	0.07	0.07	0.07	-0.24	-0.24	-0.56	-0.56
GEORGE	0.07	0.69	0.38	0.07	0.07	0.07	0.38	-0.24	-0.24	0.00	0.00	0.07	0.07	0.07	0.07	-0.56	-0.24	-0.56	-0.24	-0.56	-0.56	0.07	-0.24	0.07	-0.56	-0.56	-0.24
PETE	0.07	-0.24	0.69	0.38	0.38	0.07	0.38	0.69	-0.24	0.69	0.07	0.00	0.69	0.38	0.07	-0.56	0.07	0.07	-0.56	-0.24	-0.56	0.38	-0.24	0.07	-0.56	0.38	-0.24
FRED	0.38	-0.24	0.38	0.07	0.69	0.07	0.69	0.07	0.38	0.07	0.07	0.69	0.00	0.69	0.38	-0.56	0.07	0.38	-0.56	-0.24	-0.24	0.38	-0.24	-0.24	-0.24	0.07	0.07
GENE	0.38	-0.24	0.69	0.07	-0.24	0.07	0.38	0.07	0.07	0.07	0.07	0.38	0.69	0.00	0.38	-0.24	-0.24	0.07	-0.24	-0.24	-0.56	0.07	-0.56	-0.24	0.07	-0.24	0.07
RALPH	0.69	0.38	0.07	0.38	0.69	0.69	0.07	0.07	0.38	0.07	0.07	0.07	0.38	0.38	0.00	0.07	0.38	0.38	0.07	0.07	0.07	0.07	0.07	0.38	0.07	0.07	0.07
PHIL	-0.24	0.07	-0.24	-0.56	-0.24	-0.56	0.07	-0.24	-0.24	-0.24	-0.56	-0.56	-0.24	0.07	0.07	0.00	0.07	0.38	0.69	0.69	0.07	0.07	0.07	0.07	1.00	0.07	0.38
IKE	0.07	0.07	0.07	0.38	0.69	-0.24	-0.56	-0.24	0.38	-0.24	-0.24	0.07	0.07	-0.24	0.38	0.07	0.00	0.38	0.07	0.38	0.07	0.07	-0.24	0.07	0.07	0.07	0.07
NICK	0.07	-0.24	-0.24	-0.24	0.07	-0.56	0.07	-0.56	0.38	-0.56	-0.56	0.07	0.38	0.07	0.38	0.38	0.38	0.00	0.07	0.07	0.07	0.38	0.07	0.38	0.38	0.69	0.69
DON	-0.56	0.38	-0.24	-0.56	-0.24	-0.24	-0.24	0.07	-0.24	0.07	-0.24	-0.56	-0.56	-0.24	-0.24	0.69	0.07	0.07	0.00	0.69	0.38	0.38	0.38	0.69	0.69	0.07	0.07
NED	-0.56	0.07	0.07	-0.24	0.07	-0.56	-0.24	-0.24	0.07	-0.24	-0.56	-0.24	-0.24	-0.24	-0.24	0.69	0.38	0.07	0.69	0.00	0.38	-0.24	0.07	0.07	0.07	0.38	0.07
KARL	-0.24	-0.24	-0.24	0.07	-0.24	-0.24	0.07	-0.24	0.07	-0.24	-0.56	0.07	-0.24	-0.56	-0.56	0.07	0.07	0.07	0.38	0.38	0.00	0.07	0.69	0.07	0.07	0.38	0.07
KEN	-0.24	-0.24	0.07	-0.24	-0.24	0.07	0.07	0.07	-0.24	0.07	0.07	0.38	0.38	0.07	-0.24	0.07	-0.24	0.07	0.38	-0.24	0.07	0.00	0.38	0.38	0.00	0.38	0.07
EARL	-0.24	0.07	-0.56	-0.24	-0.56	-0.24	0.07	0.07	0.07	0.07	-0.24	-0.24	-0.24	-0.56	-0.56	0.07	0.07	0.38	0.07	0.07	0.69	0.38	0.00	0.38	0.07	0.07	0.07
RICK	-0.56	0.07	-0.56	-0.56	-0.24	0.07	-0.24	-0.24	-0.24	-0.24	-0.56	-0.56	-0.24	-0.56	-0.24	0.07	0.07	0.38	0.38	0.07	0.07	0.38	0.38	0.00	0.07	0.38	0.69
OL	-0.24	0.07	-0.24	-0.56	-0.24	-0.56	0.07	-0.24	-0.24	-0.24	0.07	0.38	-0.24	-0.24	0.07	1.00	-0.24	0.38	0.69	0.38	0.07	0.00	0.38	0.00	0.00	0.38	0.38
NEAL	-0.24	-0.24	0.07	0.07	0.38	-0.56	-0.24	-0.56	0.69	-0.56	-0.56	0.38	0.07	0.07	0.07	0.07	0.07	0.69	0.07	0.38	0.38	0.38	0.07	0.38	0.07	0.00	0.38
DAVE	-0.24	-0.24	-0.56	-0.56	-0.24	-0.24	0.07	-0.56	0.07	-0.56	-0.24	-0.24	0.07	0.07	0.07	0.38	0.07	0.69	0.07	0.07	0.07	0.07	0.07	0.69	0.38	0.38	0.00

Table 6.7 The PP weights matrix

[PP]	ART	AL	SAM	CLYDE	MIKE	JIM	GREG	JOHN	DOUG	LANCE	GEORGE	PETE	FRED	GENE	RALPH	PHIL	IKE	NICK	DON	NED	KARL	KEN	EARL	RICK	OL	NEAL	DAVE
ART	0.00	-0.01	-0.01	0.13	0.05	-0.01	-0.01	-0.01	-0.01	-0.01	-0.01	-0.01	0.05	0.05	0.13	-0.08	-0.01	-0.01	-0.52	-0.52	-0.08	-0.08	-0.08	-0.52	-0.08	-0.08	-0.08
AL	-0.01	0.00	-0.08	-0.01	0.05	0.05	-0.01	0.13	0.13	0.13	0.05	-0.08	-0.08	-0.08	0.05	-0.01	-0.01	-0.08	0.05	-0.01	-0.08	-0.08	-0.01	-0.01	-0.01	-0.08	-0.08
SAM	-0.01	-0.08	0.00	0.05	0.05	-0.01	-0.01	-0.01	0.05	-0.01	-0.01	0.13	0.05	0.13	-0.01	-0.08	-0.01	-0.08	-0.08	-0.01	-0.08	-0.01	-0.52	-0.52	-0.08	-0.01	-0.52
CLYDE	0.13	-0.01	0.05	0.00	0.13	-0.01	-0.08	0.05	-0.01	-0.01	-0.01	0.05	-0.01	0.05	-0.01	-0.52	-0.01	-0.08	-0.52	-0.08	-0.01	-0.08	-0.08	-0.52	-0.08	-0.01	-0.52
MIKE	0.05	0.05	0.05	0.13	0.00	-0.01	-0.08	-0.01	-0.01	-0.01	-0.01	0.05	0.13	-0.01	0.13	-0.08	0.05	-0.08	-0.08	-0.01	-0.08	-0.08	-0.08	-0.52	-0.52	0.05	-0.08
JIM	-0.01	0.05	-0.01	-0.01	-0.01	0.00	-0.01	0.13	-0.08	-0.01	-0.01	0.05	-0.01	0.00	-0.01	-0.52	0.13	-0.01	-0.08	-0.52	-0.08	-0.01	-0.52	-0.08	-0.08	0.05	-0.08
GREG	-0.01	0.05	-0.01	-0.08	-0.08	-0.01	0.00	0.05	0.05	-0.01	0.97	-0.01	0.13	0.05	-0.01	-0.01	-0.52	0.05	-0.08	-0.08	-0.52	-0.01	-0.08	-0.01	-0.52	-0.52	-0.01
JOHN	-0.01	0.13	-0.01	0.05	-0.01	0.13	0.05	0.00	-0.01	0.05	0.13	-0.01	-0.01	-0.01	-0.08	-0.08	-0.08	-0.52	-0.08	-0.08	-0.08	-0.01	-0.01	-0.08	-0.52	-0.08	-0.01
DOUG	-0.01	0.13	0.05	-0.01	-0.01	-0.08	0.05	-0.08	0.00	0.97	0.13	0.13	0.05	-0.01	0.05	-0.08	0.05	0.05	-0.08	-0.01	-0.01	-0.08	-0.01	-0.08	-0.08	0.13	-0.01
LANCE	-0.01	0.13	-0.01	-0.01	-0.01	-0.01	-0.01	0.05	0.97	0.00	0.00	0.13	-0.01	-0.01	-0.01	-0.52	-0.08	-0.52	-0.01	-0.08	-0.01	-0.01	-0.01	-0.08	-0.08	-0.52	-0.52
GEORGE	-0.01	0.05	-0.01	-0.01	-0.01	-0.01	0.97	0.13	0.13	0.00	0.00	-0.01	-0.01	-0.01	-0.01	-0.52	-0.08	-0.52	-0.08	-0.52	-0.52	-0.01	-0.08	-0.01	-0.52	-0.52	-0.08
PETE	-0.01	-0.08	0.13	0.05	0.05	0.05	-0.01	-0.01	0.13	0.13	-0.01	0.00	0.13	0.05	-0.01	-0.52	-0.01	-0.01	-0.52	-0.08	-0.52	0.05	-0.08	-0.08	-0.52	0.05	-0.08
FRED	0.05	-0.08	0.05	-0.01	0.13	-0.01	0.13	-0.01	0.05	-0.01	-0.01	0.13	0.00	0.13	0.05	-0.08	-0.08	0.05	-0.52	-0.52	-0.52	0.05	-0.08	0.05	-0.08	-0.01	-0.01
GENE	0.05	-0.08	0.13	0.05	-0.01	0.00	0.05	-0.01	-0.01	-0.01	-0.01	0.05	0.13	0.00	0.05	-0.01	0.00	-0.01	-0.52	-0.08	-0.52	-0.01	-0.52	-0.01	-0.01	-0.01	-0.08
RALPH	0.13	0.05	-0.01	-0.01	0.13	-0.01	-0.01	-0.08	0.05	-0.01	-0.01	-0.01	0.05	0.05	0.00	-0.01	0.05	0.05	0.05	-0.08	-0.52	-0.01	-0.52	-0.08	-0.01	-0.01	-0.01
PHIL	-0.08	-0.01	-0.08	-0.52	-0.08	-0.52	-0.01	-0.08	-0.08	-0.52	-0.52	-0.52	-0.08	-0.01	-0.01	0.00	-0.01	0.05	0.13	0.13	-0.01	-0.08	-0.08	-0.01	0.97	-0.01	0.05
IKE	-0.01	-0.01	-0.01	-0.01	0.05	0.13	-0.52	-0.08	0.05	-0.08	-0.08	-0.01	-0.08	0.00	0.05	-0.01	0.00	0.05	-0.01	0.05	-0.01	-0.01	-0.08	-0.01	-0.01	0.13	-0.01
NICK	-0.01	-0.08	-0.08	-0.08	-0.08	-0.01	0.05	-0.52	0.05	-0.52	-0.52	-0.01	0.05	-0.01	0.05	0.05	0.05	0.00	-0.01	-0.01	-0.01	0.05	-0.01	0.05	0.05	0.13	0.13
DON	-0.52	0.05	-0.08	-0.52	-0.08	-0.08	-0.08	-0.08	-0.08	-0.01	-0.08	-0.52	-0.52	-0.52	-0.08	0.13	-0.01	-0.01	0.00	0.13	-0.01	-0.01	0.05	0.05	0.13	-0.01	-0.01
NED	-0.52	-0.01	-0.01	-0.08	-0.01	-0.52	-0.08	-0.08	-0.01	-0.08	-0.52	-0.08	-0.52	-0.08	-0.08	0.13	0.05	-0.01	0.13	0.00	0.05	0.05	-0.01	-0.01	0.13	0.05	-0.01
KARL	-0.08	-0.08	-0.08	-0.01	-0.08	-0.08	-0.52	-0.08	-0.01	-0.01	-0.52	-0.52	-0.52	-0.52	-0.52	-0.01	-0.01	-0.01	-0.01	0.05	0.00	-0.01	0.13	-0.01	-0.01	0.05	-0.01
KEN	-0.08	-0.08	-0.01	-0.08	-0.08	-0.01	-0.01	-0.01	-0.08	-0.01	-0.01	0.05	0.05	-0.01	-0.01	-0.08	-0.01	0.05	-0.01	0.05	-0.01	0.00	0.05	0.05	-0.08	0.05	-0.01
EARL	-0.08	-0.01	-0.52	-0.08	-0.08	-0.52	-0.08	-0.01	-0.01	-0.01	-0.08	-0.08	-0.08	-0.52	-0.52	-0.08	-0.08	-0.01	0.05	-0.01	0.13	0.05	0.00	0.05	-0.01	-0.01	-0.01
RICK	-0.52	-0.01	-0.52	-0.52	-0.52	-0.08	-0.01	-0.08	-0.08	-0.08	-0.01	-0.08	0.05	-0.01	-0.08	-0.01	-0.01	0.05	0.05	-0.01	-0.01	0.05	0.05	0.00	-0.01	0.05	0.13
OL	-0.08	-0.01	-0.08	-0.08	-0.52	-0.08	-0.52	-0.52	-0.08	-0.08	-0.52	-0.52	-0.08	-0.01	-0.01	0.97	-0.01	0.05	0.13	0.13	-0.01	-0.08	-0.01	-0.01	0.00	0.00	0.05
NEAL	-0.08	-0.08	-0.01	-0.01	0.05	0.05	-0.52	-0.08	0.13	-0.52	-0.52	0.05	-0.01	-0.01	-0.01	-0.01	0.13	0.13	-0.01	0.05	0.05	0.05	-0.01	0.05	-0.01	0.00	0.05
DAVE	-0.08	-0.08	-0.52	-0.52	-0.08	-0.08	-0.01	-0.01	-0.01	-0.52	-0.08	-0.08	-0.01	-0.08	-0.01	0.05	-0.01	0.13	-0.01	-0.01	-0.01	-0.01	-0.01	0.13	0.05	0.05	0.00

Table 6.8 The ACS–ULA weights matrix

[ACS]	ART	AL	SAM	CLYDE	MIKE	JIM	GREG	JOHN	DOUG	LANCE	GEORGE	PETE	FRED	GENE	RALPH	PHIL	IKE	NICK	DON	NED	KARL	KEN	EARL	RICK	OL	NEAL	DAVE
ART	0.00	1.00	1.00	1.00	1.00	1.00	1.00	1.00	1.00	1.00	1.00	1.00	1.00	1.00	1.00	-1.00	-1.00	-1.00	-1.00	-1.00	-1.00	-1.00	-1.00	-1.00	-1.00	-1.00	-1.00
AL	1.00	0.00	1.00	1.00	1.00	1.00	1.00	1.00	1.00	1.00	1.00	1.00	1.00	1.00	1.00	-1.00	-1.00	-1.00	-1.00	-1.00	-1.00	-1.00	-1.00	-1.00	-1.00	-1.00	-1.00
SAM	1.00	1.00	0.00	1.00	1.00	1.00	1.00	1.00	1.00	1.00	1.00	1.00	1.00	1.00	1.00	-1.00	-1.00	-1.00	-1.00	-1.00	-1.00	-1.00	-1.00	-1.00	-1.00	-1.00	-1.00
CLYDE	1.00	1.00	1.00	0.00	0.00	1.00	1.00	1.00	1.00	1.00	1.00	1.00	1.00	1.00	1.00	-1.00	-1.00	-1.00	-1.00	-1.00	-1.00	-1.00	-1.00	-1.00	-1.00	-1.00	-1.00
MIKE	1.00	1.00	1.00	0.00	0.00	1.00	1.00	1.00	1.00	1.00	1.00	1.00	1.00	1.00	1.00	-1.00	-1.00	-1.00	-1.00	-1.00	-1.00	-1.00	-1.00	-1.00	-1.00	-1.00	-1.00
JIM	1.00	1.00	1.00	1.00	1.00	0.00	1.00	1.00	1.00	1.00	1.00	1.00	1.00	1.00	1.00	-1.00	-1.00	-1.00	-1.00	-1.00	-1.00	-1.00	-1.00	-1.00	-1.00	-1.00	-1.00
GREG	1.00	1.00	1.00	1.00	1.00	1.00	0.00	0.00	1.00	1.00	1.00	1.00	1.00	1.00	1.00	-1.00	-1.00	-1.00	-1.00	-1.00	-1.00	-1.00	-1.00	-1.00	-1.00	-1.00	-1.00
JOHN	1.00	1.00	1.00	1.00	1.00	1.00	0.00	0.00	1.00	1.00	1.00	1.00	1.00	1.00	1.00	-1.00	-1.00	-1.00	-1.00	-1.00	-1.00	-1.00	-1.00	-1.00	-1.00	-1.00	-1.00
DOUG	1.00	1.00	1.00	1.00	1.00	1.00	1.00	1.00	0.00	0.00	1.00	1.00	1.00	1.00	1.00	-1.00	-1.00	-1.00	-1.00	-1.00	-1.00	-1.00	-1.00	-1.00	-1.00	-1.00	-1.00
LANCE	1.00	1.00	1.00	1.00	1.00	1.00	1.00	1.00	0.00	0.00	1.00	1.00	1.00	1.00	1.00	-1.00	-1.00	-1.00	-1.00	-1.00	-1.00	-1.00	-1.00	-1.00	-1.00	-1.00	-1.00
GEORGE	1.00	1.00	1.00	1.00	1.00	1.00	1.00	1.00	1.00	1.00	0.00	0.00	1.00	1.00	1.00	-1.00	-1.00	-1.00	-1.00	-1.00	-1.00	-1.00	-1.00	-1.00	-1.00	-1.00	-1.00
PETE	1.00	1.00	1.00	1.00	1.00	1.00	1.00	1.00	1.00	1.00	0.00	0.00	1.00	1.00	1.00	-1.00	-1.00	-1.00	-1.00	-1.00	-1.00	-1.00	-1.00	-1.00	-1.00	-1.00	-1.00
FRED	1.00	1.00	1.00	1.00	1.00	1.00	1.00	1.00	1.00	1.00	1.00	1.00	0.00	0.00	1.00	-1.00	-1.00	-1.00	-1.00	-1.00	-1.00	-1.00	-1.00	-1.00	-1.00	-1.00	-1.00
GENE	1.00	1.00	1.00	1.00	1.00	1.00	1.00	1.00	1.00	1.00	1.00	1.00	0.00	0.00	1.00	-1.00	-1.00	-1.00	-1.00	-1.00	-1.00	-1.00	-1.00	-1.00	-1.00	-1.00	-1.00
RALPH	1.00	1.00	1.00	1.00	1.00	1.00	1.00	1.00	1.00	1.00	1.00	1.00	1.00	1.00	0.00	-1.00	-1.00	-1.00	-1.00	-1.00	-1.00	-1.00	-1.00	-1.00	-1.00	-1.00	-1.00
PHIL	-1.00	-1.00	-1.00	-1.00	-1.00	-1.00	-1.00	-1.00	-1.00	-1.00	-1.00	-1.00	-1.00	-1.00	-1.00	0.00	0.00	1.00	1.00	1.00	1.00	1.00	1.00	1.00	1.00	1.00	1.00
IKE	-1.00	-1.00	-1.00	-1.00	-1.00	-1.00	-1.00	-1.00	-1.00	-1.00	-1.00	-1.00	-1.00	-1.00	-1.00	0.00	0.00	1.00	1.00	1.00	1.00	1.00	1.00	1.00	1.00	1.00	1.00
NICK	-1.00	-1.00	-1.00	-1.00	-1.00	-1.00	-1.00	-1.00	-1.00	-1.00	-1.00	-1.00	-1.00	-1.00	-1.00	1.00	1.00	0.00	0.00	1.00	1.00	1.00	1.00	1.00	1.00	1.00	1.00
DON	-1.00	-1.00	-1.00	-1.00	-1.00	-1.00	-1.00	-1.00	-1.00	-1.00	-1.00	-1.00	-1.00	-1.00	-1.00	1.00	1.00	0.00	0.00	1.00	1.00	1.00	1.00	1.00	1.00	1.00	1.00
NED	-1.00	-1.00	-1.00	-1.00	-1.00	-1.00	-1.00	-1.00	-1.00	-1.00	-1.00	-1.00	-1.00	-1.00	-1.00	1.00	1.00	1.00	1.00	0.00	0.00	1.00	1.00	1.00	1.00	1.00	1.00
KARL	-1.00	-1.00	-1.00	-1.00	-1.00	-1.00	-1.00	-1.00	-1.00	-1.00	-1.00	-1.00	-1.00	-1.00	-1.00	1.00	1.00	1.00	1.00	0.00	0.00	1.00	1.00	1.00	1.00	1.00	1.00
KEN	-1.00	-1.00	-1.00	-1.00	-1.00	-1.00	-1.00	-1.00	-1.00	-1.00	-1.00	-1.00	-1.00	-1.00	-1.00	1.00	1.00	1.00	1.00	1.00	1.00	0.00	0.00	1.00	1.00	1.00	1.00
EARL	-1.00	-1.00	-1.00	-1.00	-1.00	-1.00	-1.00	-1.00	-1.00	-1.00	-1.00	-1.00	-1.00	-1.00	-1.00	1.00	1.00	1.00	1.00	1.00	1.00	0.00	0.00	1.00	1.00	1.00	1.00
RICK	-1.00	-1.00	-1.00	-1.00	-1.00	-1.00	-1.00	-1.00	-1.00	-1.00	-1.00	-1.00	-1.00	-1.00	-1.00	1.00	1.00	1.00	1.00	1.00	1.00	1.00	1.00	0.00	0.00	1.00	1.00
OL	-1.00	-1.00	-1.00	-1.00	-1.00	-1.00	-1.00	-1.00	-1.00	-1.00	-1.00	-1.00	-1.00	-1.00	-1.00	1.00	1.00	1.00	1.00	1.00	1.00	1.00	1.00	0.00	0.00	1.00	1.00
NEAL	-1.00	-1.00	-1.00	-1.00	-1.00	-1.00	-1.00	-1.00	-1.00	-1.00	-1.00	-1.00	-1.00	-1.00	-1.00	1.00	1.00	1.00	1.00	1.00	1.00	1.00	1.00	1.00	1.00	0.00	1.00
DAVE	-1.00	-1.00	-1.00	-1.00	-1.00	-1.00	-1.00	-1.00	-1.00	-1.00	-1.00	-1.00	-1.00	-1.00	-1.00	1.00	1.00	1.00	1.00	1.00	1.00	1.00	1.00	1.00	1.00	1.00	0.00

Table 6.9 Linear correlation before and after the application of ULA and meta correlation between the two matrices

ULA dataset

correlation	Jet	Sharks	20's	30's	40's	JH	COL	HS	Single	Married	Divorced	Pusher	Bookie	Burglar
Jet	0.0000	-1.0000	0.5316	-0.4807	-0.0466	0.5316	-0.2390	-0.3202	0.2652	-0.2401	-0.0466	0.0000	0.0000	0.0000
Sharks	-1.0000	0.0000	-0.5316	0.4807	0.0466	-0.5316	0.2390	0.3202	-0.2652	0.2401	0.0466	0.0000	0.0000	0.0000
20's	0.5316	-0.5316	0.0000	-0.7391	-0.3198	0.0471	-0.0410	-0.0116	0.0284	-0.1118	0.1119	-0.0542	-0.2169	0.2712
30's	-0.4807	0.4807	-0.7391	0.0000	-0.4019	-0.1251	0.1981	-0.0447	-0.0385	0.0284	0.0155	0.1048	0.1048	-0.2097
40's	-0.0466	0.0466	-0.3198	-0.4019	0.0000	0.1119	-0.2229	0.0786	0.0155	0.1119	-0.1739	-0.0737	0.1474	-0.0737
JH	0.5316	-0.5316	0.0471	-0.1251	0.1119	0.0000	-0.4100	-0.6359	0.0284	-0.1118	0.1119	-0.2169	-0.0542	0.2712
COL	-0.2390	0.2390	-0.0410	0.1981	-0.2229	-0.4100	0.0000	-0.4432	-0.1585	0.3280	-0.2229	0.1890	0.0000	-0.1890
HS	-0.3202	0.3202	-0.0116	-0.0447	0.0786	-0.6359	-0.4432	0.0000	0.1062	-0.1677	0.0786	0.0533	0.0533	-0.1066
Single	0.2652	-0.2652	0.0284	-0.0385	0.0155	0.0284	-0.1585	0.1062	0.0000	-0.7391	-0.4019	0.1048	0.4193	-0.5241
Married	-0.2401	0.2401	-0.1118	0.0284	0.1119	-0.1118	0.3280	-0.1677	-0.7391	0.0000	-0.3198	-0.0542	-0.2169	0.2712
Divorced	-0.0466	0.0466	0.1119	0.0155	-0.1739	0.1119	-0.2229	0.0786	-0.4019	-0.3198	0.0000	-0.0737	-0.2949	0.3686
Pusher	0.0000	0.0000	-0.0542	0.1048	-0.0737	-0.2169	0.1890	0.0533	0.1048	-0.0542	-0.0737	0.0000	-0.5000	-0.5000
Bookie	0.0000	0.0000	-0.2169	0.1048	0.1474	-0.0542	0.0000	0.0533	0.4193	-0.2169	-0.2949	-0.5000	0.0000	-0.5000
Burglar	0.0000	0.0000	0.2712	-0.2097	-0.0737	0.2712	-0.1890	-0.1066	-0.5241	0.2712	0.3686	-0.5000	-0.5000	0.0000

ULA dataset

Correlation	Jet	Sharks	20's	30's	40's	JH	COL	HS	Single	Married	Divorced	Pusher	Bookie	Burglar
Jet	0.0000	-0.9880	0.9862	-0.9751	0.2069	0.9872	-0.9534	-0.9587	0.6094	-0.9188	0.4911	-0.9225	0.1566	0.5570
Sharks	-0.9880	0.0000	-0.9810	0.9869	-0.2075	-0.9845	0.9663	0.9646	-0.6008	0.9301	-0.5070	0.9294	-0.1400	-0.5660
20's	0.9862	-0.9810	0.0000	-0.9747	0.1893	0.9850	-0.9547	-0.9614	0.5890	-0.9133	0.5223	-0.9306	0.1357	0.5858
30's	-0.9751	0.9869	-0.9747	0.0000	-0.2180	-0.9787	0.9783	0.9742	-0.5766	0.9320	-0.5235	0.9471	-0.1216	-0.5890
40's	0.2069	-0.2075	0.1893	-0.2180	0.0000	0.2106	-0.2171	-0.2038	0.5557	-0.2354	-0.3742	-0.1956	0.7706	-0.2950
JH	0.9872	-0.9845	0.9850	-0.9787	0.2106	0.0000	-0.9655	-0.9722	0.5928	-0.9203	0.5187	-0.9402	0.1425	0.5824

(continued)

Table 6.9 (continued)

ULA dataset correlation	Jet	Sharks	20s	30s	40s	JH	COL	HS	Single	Married	Divorced	Pusher	Bookie	Burglar
COL	-0.9534	0.9663	-0.9547	0.9783	-0.2171	-0.9655	0.0000	0.9671	-0.5702	0.9408	-0.5430	0.9515	-0.1136	-0.5888
HS	-0.9587	0.9646	-0.9614	0.9742	-0.2038	-0.9722	0.9671	0.0000	-0.5607	0.9314	-0.5583	0.9683	-0.1252	-0.6169
Single	0.6094	-0.6008	0.5890	-0.5766	0.5557	0.5928	-0.5702	-0.5607	0.0000	-0.7006	-0.0076	-0.5368	0.7470	0.0356
Married	-0.9188	0.9301	-0.9133	0.9320	-0.2354	-0.9203	0.9408	0.9314	-0.7006	0.0000	-0.4485	0.9195	-0.2454	-0.4760
Divorced	0.4911	-0.5070	0.5223	-0.5235	-0.3742	0.5187	-0.5430	-0.5583	-0.0076	-0.4485	0.0000	-0.5817	-0.3855	0.9450
Pusher	-0.9225	0.9294	-0.9306	0.9471	-0.1956	-0.9402	0.9515	0.9683	-0.5368	0.9195	-0.5817	0.0000	-0.1279	-0.6470
Bookie	0.1566	-0.1400	0.1357	-0.1216	0.7706	0.1425	-0.1136	-0.1252	0.7470	-0.2454	-0.3855	-0.1279	0.0000	-0.3439
Burglar	0.5570	-0.5660	0.5858	-0.5890	-0.2950	0.5824	-0.5888	-0.6169	0.0356	-0.4760	0.9450	-0.6470	-0.3439	1.0000
Meta correlation (Source-ULA)	0.8042	0.8179	0.6238	0.7074	0.3494	0.8279	0.6019	0.2299	0.5072	0.4736	0.5496	0.4325	0.7017	0.5742

- Some Meta correlations are very low (40s, HS, Single, Married, Divorced, Pusher, Burglar): This means that the amount of information generated by ULA is substantial.

From the point of view of the generation of new information, two different kinds of changes can be noted:

- A *quantitative* modification occurs when a variable reinforces its relationship of excitation or of inhibition with some others. For example, with reference to our database "JH" and "20s" increase their positive relationship with "Jet" and their negative relation with "Sharks". On the opposite side, "HS", "College" and "30s" increase their solidarity with "Sharks" and their opposition to "Jets";
- A *qualitative* modification occurs when a variable changes the nature of its relationships with the others. In the case of our database, this is evident with the variables "40s" and "Divorced." In the original matrix, their relationship with both "Jets" and "Sharks" is weak and slightly oriented toward the "Sharks", whereas in the final matrix both variables have become completely "Jet" oriented: This means that, whereas by looking at the global structure of the database the variables "40s" and "Divorced" play an ambiguous role, for the point of view of the inherent structure of the association of variables from the optimal sorting point of view, their presence tends to be more meaningful for the identification of Jets (where their presence is more atypical, being the majority of Jets relatively young and single) than of Sharks.

Another interesting qualitative change brought about by ULA concerns the variables "College" and "HS". These variables are orthogonal in the source dataset. The presence of any of them excludes the other one. Consequently, from the viewpoint of a simple linear correlation analysis (original matrix), their relationship looks like strong mutual inhibition (-0.4432). But, according to ULA, this is a misleading conclusion. In fact, both the variables "HS" and "College" characterize in a strong way the variable "Sharks" and both inhibit the variable "Jets". That is to say: When a "Sharks" member is not "HS", he is probably "College", and vice versa. So, despite their orthogonal relationship, these two variables support a similar prototype.

The same pattern of non linear connection is found by ULA among the last three variables: "Pusher", "Bookie" and "Burglar". In the source dataset, they are orthogonal to each other, and their weighted frequencies are exactly the same in the two main classes ("Jets" and "Sharks"). In fact, in the first correlation matrix their relationships with "Jets" and "Sharks" are null. But, from the ULA point of view, "Pusher" is strongly "Sharks" oriented, "Burglar" is relatively more "Jet" oriented, and "Bookie" presents a fuzzy positive membership with "Jets". Looking at Carnot maps in detail (Tables 6.4 and 6.5), this pattern becomes evident.

In Fig. 6.2, we present, in detail for each one of the 14 variables, the meta-correlation between the correlation in the source dataset and the correlation in the dataset generated by ULA.

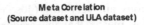

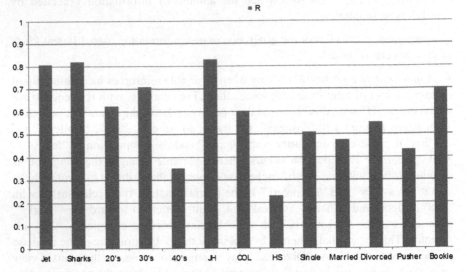

Fig. 6.2 Meta-correlation between the source dataset and ULA

6.8 Optimal Filtering

At this point, we are ready to translate the ULA methodology into actual optimal sorting inferences and to apply a benchmark analysis. To appreciate the impact of the ULA methodology, we preliminarily apply an optimal filtering technique to the more conventional approaches (LC and PP) showing what kind of "minimal spanning tree" can be obtained by selecting the optimal tree (in terms of relative distances as measured on the basis of the corresponding weights matrix), and discussing the informational sorting that they operate on the West Side Story database. To this purpose we make use of the Population algorithm (Massini et al. 2010), which operates on the basis of the minimization of a certain measure of dissimilarity between the matrix of the original distances and that of the mapped distances.

In the case of LC, presented in Fig. 6.3, the sorting is imperfect. Specifically, two Sharks, Ike and Ken, get assigned to the Jets. The border between the two gangs is found in Doug, a Jet. Doug presents in fact a few anomalous characters as a Jet: he is in the 30s and has a High School education, while at the same time being single, so he is a proper hybrid of "Jetness" and "Sharkness". The two incorrectly attributed individuals, Ken and Ike, are again two hybrids: Ike is in the 30s but has a Junior High School education and is single, whereas Ken is in the 20s and is single, albeit having a High School education. Thus, LC analysis seems to be very sensitive to misattributing Sharks when they are single and presents another

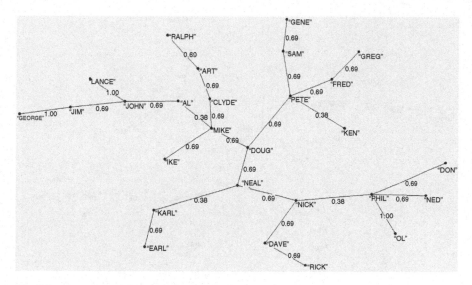

Fig. 6.3 MST resulting from LC analysis

characteristic that is distinctive of "Jetness". This finding reflects the ambiguity in the role of the single versus married family status that emerged in the analysis *before* the application of ULA in Sect. 6.2 above.

In the case of PP, sorting is again imperfect and actually replicates the outcome for the LC methodology, generating a topologically equivalent MST (despite the corresponding weight matrices are different): Doug is the "border" individual and Ken and Ike are attributed to the Jets, and also the structure of the branches linking all other individuals are the same, as shown in Fig. 6.4. Thus, the two methods turn out to replicate the same errors and to manifest the same bias in terms of improper use of the informational content of variables for sorting purposes.

What happens once the two weights matrices are fed into the ACS algorithm and ULA analysis is carried out? The application of the ULA algorithm generates a new database where each record has 27-dimensions. We make use of the Population (Massini et al. 2010) algorithm to project such trajectories in a lower-dimensional space, thus generating the corresponding MST and the 3D rendition of the UL trajectories. As to the MST, a radically different result emerges, as depicted in Fig. 6.5.

In the case of ULA, not only are all individuals properly attributed, but the way in which the underlying structure of associations is depicted is much different, and in particular, radically different from the one emerging from *both of the input* weight matrices. The ACS-ULA methodology has thoroughly re-organized the information provided by LC and PP into an entirely original, and accurate, scheme. Again, Doug is the borderline individual. Unlike the LC and PP representations, where the distribution of weights across the graph measured the connections between individuals with varying strengths, in the ACS-ULA MST all individuals

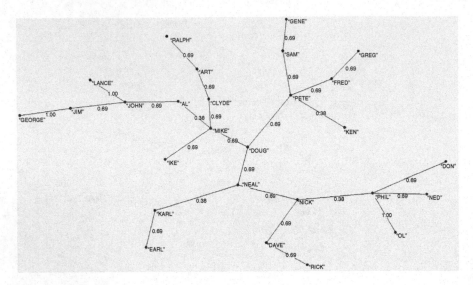

Fig. 6.4 MST resulting from PP analysis

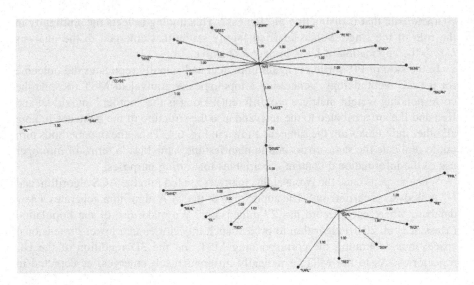

Fig. 6.5 MST resulting from ACS-ULA analysis

belonging to the same gang are connected with maximum strength (+1), whereas the connection between the borderline individuals separating the two gangs (Doug and Ken) is the only one that presents a maximal *negative* strength (−1). That is to say, ACS-ULA has perfectly *understood* the structure of affiliation that was embodied in the data and has mapped it without uncertainties.

As to the structure of the connections between individuals, all members of a gang are presented as leaves of a small number of "hub" individuals around whom specific membership sub-classes are organized. First of all, Jets present a "double border" structure where another individual, Lance, sits as a gateway to "Sharkness" in a subordinate position with respect to Doug. Lance is in the 20s and has a Junior High School education, but is married, i.e., he presents the complementary characteristic with respect to the "anomalous" Sharks Ike and Ken. All remaining Jets other than Doug and Lance, however, can be seen as "leaves" of Art, who has Junior High education and is single, but is in the 40s: A highly anomalous feature for a Jet – a trait that is shared only by another Jet, Clyde, and which, as we know from the analysis of Sect. 6.2, is taken by ACS-ULA as a highly meaningful carrier of optimal sorting information for "Jetness". As to Sharks, they are organized around two hubs, namely, Ken and Earl. Interestingly, Ken is one of the two "anomalous" Sharks who was mistakenly included in the Jets according to LC and PP approaches: Thus, his failed attribution is not simply a mistake, but causes a substantial misunderstanding of the true structure of association among variables on the Sharks side. Moreover, Ken is directly linked with Doug, and thus represents the true borderline Shark individual. Unlike the other hubs, Earl, who is directly linked to Ken, is an almost a perfect Shark: He is married and has a High School education, his only anomaly being age. He is in his 40s (which is, however, less anomalous than for Art, provided that the representative age for Jets is in the 20s rather than in the 30s).

Notice that in the LC and PP MSTs, no hubness structure emerged, and the graph topology for the Jet and the Shark groups was pretty much similar. In the ULA representation, to the contrary, hubness properties are essential to understand the association patterns within groups, and the graph topologies for the two groups are substantially different – a single hub structure for Jets with a "thick border", a dual hub structure for Sharks with a "thin border".

The discriminating capacity of ULA can be appreciated even better by means of a 3D depiction of the dynamics of Universe Lines that is presented in Figs. 6.6, 6.7, and 6.8. The figures show clearly how, in the environment space, the two populations are strongly entangled as to the mix of individual characteristics, but as the successive iterations of ULA made the informational roles of variables for sorting purposes clearer, they tend to be separated increasingly well, landing in totally different regions of the space. In Fig. 6.6, we see how the UL trajectories clearly separate Jets from Sharks individuals in the 3D space, with the exception of one individual whose trajectory remains eccentric with respect to those of the main groups, but however firmly placed on the Jets side. Figure 6.7 reveals that this individual is actually Doug, namely, the borderline one that separates the two groups. Figure 6.8, finally, reveals that the two anomalous Shark individuals, namely Ike and Ken, initially follow a trajectory that is slightly anomalous with respect to the group (that is what leads non-dynamic algorithms such as LC and PP into error), but that as iterations proceed, they get fully absorbed into the proper group – as the iterations sort out variables with increasing efficiency, individuals like Ike and Ken appear less anomalous and coherent with the complex typical characterization of their group.

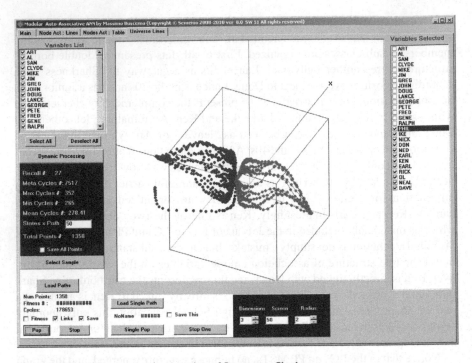

Fig. 6.6 UL in 3D space: optimal sorting of Jets versus Sharks

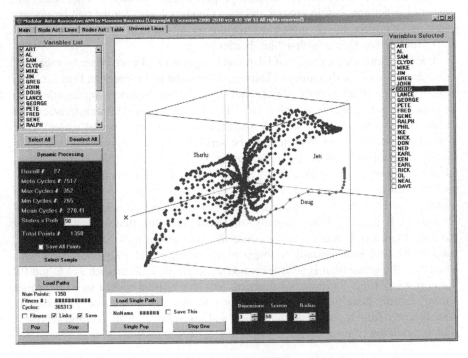

Fig. 6.7 UL in 3D space: the anomalous trajectory of Doug

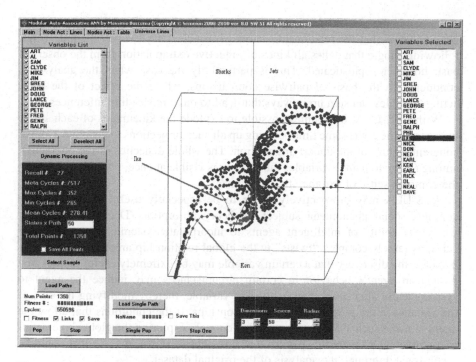

Fig. 6.8 UL in 3D space: the transient anomalies of Ike and Ken

This sharp visual separability of gangs in the 3D space induced by ULA reflects the informational gain that is produced through the enhancement in the classification of variables for the purpose of optimal discrimination as explained in Sect. 6.2. As the iterations of ULA proceed, the association of certain combinations of variables with gang affiliation becomes clearer, and even variables that were previously uninformative – in that they are evenly distributed across gangs (such as criminal profession) – or variables that played an ambiguous role, now become more and more informative as they turn out to present different *joint* discriminatory value depending on the gang. As a consequence, individuals carrying hybrid sets of characteristics who were previously likely to be miss-assigned now acquire a higher typicality in their own group and contribute to increase the sorting efficiency of the methodology.

6.9   Conclusion

Complex patterns of association between variables for significantly non linear phenomena call for more sophisticated statistical techniques. An observational sample is like an instant photograph of a body in motion, which may in principle be compatible with many different motions, the more so the smaller the variability contained in the original dataset. In particular, what happens in a highly nonlinear phenomena is that the relationships among variables may evolve in subtle ways, following a logic that defies all kinds of inductive extrapolation from the observed past, however sophisticated. This is particularly the case when the analysis is conducted on the basis of pairwise comparisons, where the effect of the other acting variables can sum up in ways that lead to quite misleading inferences.

With ACS-ULA, it becomes possible to explode the kinematics of each single variable, while at the same time stacking up all such projections to generate a global coherent picture of the underlying motion. The whole dialectics of the interaction among (many to many) variables thus becomes visible at once, possibly subverting the common sense of extrapolative thought.

ACS-ULA may prospectively prove to be especially useful in dealing with complex social phenomena subject to theory absorption (Dacey 1976, 1979), where the ability of intelligent agents to incorporate systemic knowledge may cause extremely complex "twists" in the global relationship among variables. For instance, the discovery that a certain variable may be extremely useful to track and predict an agent's behavior in certain circumstances may induce the agent to systematically alter his responses to such variable, thereby modifying the role of the variable in the global system of relationships. If the conditions that allow the agent to make such a discovery are stable enough, exploring the global kinematics of variables will reveal this pattern that turns upside down, the one that could be reconstructed through the analysis of the original dataset.

In a sense, the ACS-ULA approach may be considered a first step toward a new "physics of global interactions", where one can analyze all kinds of context-sensitive behaviors, while being at the same time able to evaluate to what extent the phenomenon under study really calls for the ULA expanded representation or not. This naturally leads to the possibility of classifying the complexity of phenomena on the basis of the actual informational gain that is generated by the passage from the original dataset to the expanded one, for the wider the gap, the more complex the phenomenon in a highly specific sense. It would be interesting to check, through suitable meta-analyses, which kinds of phenomena fall into the same equivalence classes. This could be the starting point of a new unified approach to a truly cross-disciplinary science of complexity. We look forward to explore these exciting prospects in forthcoming publications.

References

Buscema M (2007a) Squashing theory and contractive map network. Technical paper #32, Semeion, Rome
Buscema M (2007b) A novel adapting mapping method for emergent properties discovery in data bases: experience in medical field. In: 2007 I.E. international conference on systems, man and cybernetics (SMC 2007), Montreal, 7–10 Oct
Buscema M (2009) Meta auto-associative. Semeion software # 51, Rome
Dacey R (1976) Theory absorption and the testability of economic theory. J Econ 36:247–267

Dacey R (1979) The effects of theory and data in economic prediction. Kyklos 28:407–411

Grossberg S (1976) Adaptive pattern classification and universal recording: Part I. Parallel development and coding of neural feature detectors. Biol Cybern 23:121–134

Grossberg S (1978) A theory of visual coding, memory, and development. In: Leeuwenberg J, Buffart HFJ (eds) Formal theories of visual perception. Wiley, New York

Grossberg S (1980) How does the brain build a cognitive code? Psychol Rev 87:1–51

Hinton GE, Anderson JA (eds) (1981) Parallel models of associative memory. Erlbaum, Hillsdale

Kohonen T (1995) Self-organizing maps. Springer, Berlin

Massini G, Terzi S, Buscema M (2010) A new method for multi-dimensional scaling. Proceedings of NAFIPS 2010, Ryerson University, Toronto, 12–14 July

McClelland JL (1981) Retrieving general and specific information from stored knowledge of specifics. In: Proceedings of the third annual meeting of the cognitive science society, Berkeley, CA, 19–21 Aug, pp 170–172

McClelland JL (1995) Constructive memory and memory distortions: a parallel-distributed processing approach. In: Schacter DL (ed) Memory distortion. How minds, brains, and societies reconstruct the past. Harvard University Press, Cambridge, MA, pp 69–90

McClelland JL, Rumelhart DE (1988) Explorations in parallel distributed processing. A handbook of models, programs, and exercises. MIT Press, Cambridge

Chapter 7
GUACAMOLE: A New Paradigm for Unsupervised Competitive Learning

Massimo Buscema and Pier Luigi Sacco

7.1 Introduction: Supervised Versus Unsupervised Learning – In Search of a New Paradigm

It is well known that learning is a truly complex phenomenon. There are many different forms of learning, and their codification is far from straightforward. In the AI-oriented literature (and in particular for Artificial Neural Networks (ANNs)), the basic distinction that prevails to classify the variety of possibilities at a very basic level is that between supervised and unsupervised learning. In the supervised form, learners are subject to schemes of rewards and punishments that are conditional to the achievement of an external, pre-determined target, and the level of reward/punishment depends on the level of achievement of the target subject to a certain metric. In the unsupervised form, learners are free from predetermined targets and from external enforcement, and they have to derive an endogenous, idiosyncratic categorization of the phenomenon. One can imagine several hybrid forms where either the prescription value of the target or the operation of the reward system (or both) are loosened to allow for some degree of autonomy of the learner's information acquisition and categorization strategy (Chapelle et al. 2006; Zhu and Goldberg 2009). In the case of reinforcement learning, for instance, the learner generally knows nothing about the objective of the learning process but is externally rewarded/punished depending on the actions undertaken, and it is the actual greed for reward/aversion for reinforcement that drives the whole learning process

M. Buscema (✉)
Semeion Research Center of Sciences of Communication, Via Sersale 117, Rome, Italy

Department of Mathematical and Statistical Sciences, CCMB, University of Colorado, Denver, Colorado, USA
e-mail: m.buscema@semeion.it

P.L. Sacco
Semeion Research Center of Sciences of Communication, Via Sersale 117, Rome, Italy

W.J. Tastle (ed.), *Data Mining Applications Using Artificial Adaptive Systems*,
DOI 10.1007/978-1-4614-4223-3_7, © Springer Science+Business Media New York 2013

(Sutton and Barto 1998). The literature on learning, both on the human and on the machine side, is enormous and even a sketchy review is beyond the scope of this paper. It may be useful, however, to consider some contributions that are useful to define the theoretical context of our paper.

In the human context, both supervised and unsupervised learning clearly have relevance at the social level, and it is easy to number a variety of situations and even of institutions that are especially designed to implement specific supervised and/or unsupervised forms. But as it comes to the actual mechanics of individual learning at the neural level, the issue becomes much more controversial: Though humans are sensible to external rewards and punishments and can become quite able at hitting certain targets, this does not mean that a similar dynamic can describe the actual process of cognition at the *biological* level (e.g. Damper et al. 2000). At the moment, we have no evidence of the existence of 'teacher neurons' that set targets and rewards in order to direct and optimize learning content and pace of the learning system at the neural level. On the other hand, although the distance between our understanding of human learning and actual models of machine learning is still remarkable, we begin to have examples of unsupervised machine learning processes which, with some suitable amendments, are coming to reflect part of the available evidence about human learning (Körding and König 2001). And conversely, we begin to understand what kind of learning styles tend to emerge in systems calibrated upon the performances of humans in certain unsupervised learning environments, thereby illustrating once more the multi-faceted, context-sensitive nature of unsupervised learning, and allowing for comparisons with the outcomes of supervised learning (Love 2002). One of the most interesting dimensions of unsupervised learning when compared to supervised learning, is its sensitivity to redundancy, which allows the learner (be it a human or a machine) to build up a 'typical' representation of the phenomenon and to single out effectively atypical (i.e., unexpected) incoming signals (Barlow 1989). As we shall see, expectation formation and redundancy-based detection of anomalies plays a central role in our approach to unsupervised learning.

The reason why much attention is paid to supervised forms of learning is of course that they seem to allow a much more effective and efficient learning for certain tasks where the goal is clearly fixed and the degree of accomplishment is easily measurable. On the other hand, supervised learning calls for often uneconomical and cognitively limitative pre-formatting of the learning task and data (Ko and Seo 2000) – and this is a particularly serious problem when relatively little is known about the patterns to be learned (Bailey and Elkan 1995). Unsurprisingly then, there is a proliferation of hybrid approaches where both modes are eclectically implemented from time to time according to the nature of the task (Malakooti and Raman 2000), or are subsumed into a more general, abridged or unifying approach (Nadal and Parga 1994; Xu 1994; Zhao and Liu 2007), possibly calibrated upon actual human learning processes (Gureckis and Love 2003). Also, general criteria for performance comparisons between supervised and unsupervised learning processes are being developed (Lange et al. 2002). In various cases, the unsupervised approach is introduced as a task- or method-focused counterpart to well-established supervised learning techniques, to evaluate to what extent one can still achieve satisfactory results by loosening up supervision, or to derive innovative approaches by analogies in cases where actual supervision is not possible or viable (Hansen and Larsen 1996; Japkowicz 2001; Kim et al. 2002). In some cases, the unsupervised counterpart even outperforms the supervised one – an outcome that is typically welcomed with surprise by the researchers (Yarowsky 1995; El-Yaniv and Souroujon 2001), on the basis of the implicit presumption that, in terms of learning effectiveness, learners should do better when skillfully driven toward the desired result by an external trainer. And clearly, cases for the supervised approach against unsupervised ones on the basis of accuracy and computational efficiency are not lacking (Carneiro et al. 2007).

An interesting viewpoint on this theoretical debate can be gained by noting that, from the point of view of the energy function that is being calculated by an unsupervised versus a supervised ANN, it is easy to subsume both approaches into a common framework. The energy function for a supervised ANN can be written as its Mean Square Error:

$$MSE = \frac{1}{2} \cdot \sum_{p}^{K} \sum_{i}^{N} \left(t_{p,i} - u_{p,i}\right)^2; \qquad (7.1)$$

where:

t_{pi} = the i-th target of the p-th pattern;
u_{pi} = the i-th output of the p-th pattern;
K = Number of patterns;
N = Number of outputs

whereas, traditionally, the energy minimization function in an unsupervised auto-associative neural network is represented by the following equation:

$$En = \frac{1}{2} \cdot \sum_{p}^{K} \sum_{i}^{N} \sum_{j}^{N} u_{p,i} \cdot u_{p,j} \cdot w_{i,j}; \qquad (7.2)$$

where;
$w_{i,j}$ = trained weights from input j to output i.

But if we assume that (7.1) represents the mean error of a linear perceptron, then we can develop (7.1) as follows:

$$MSE = \frac{1}{2} \cdot \sum_{p}^{K} \sum_{i}^{N} \left(t_{p,i} - u_{p,i}\right)^2 = \frac{1}{2} \cdot \sum_{p}^{K} \sum_{i}^{N} \left(t_{p,i} - \sum_{j}^{N} u_{p,j} \cdot w_{i,j}\right)^2 =$$

$$= \frac{1}{2} \cdot \sum_{p}^{K} \sum_{i}^{N} \left(t_{p,i}^2 - 2 \cdot t_{p,i} \cdot \sum_{j}^{N} u_{p,j} \cdot w_{i,j} + \left(\sum_{j}^{N} u_{p,j} \cdot w_{i,j}\right)^2\right). \qquad (7.3)$$

Setting all targets to 0, as in the case of unsupervised neural networks, we have:

$$MSE = \frac{1}{2} \cdot \sum_{p}^{K} \sum_{i}^{N} \left(\sum_{j}^{N} u_{p,j} \cdot w_{i,j} \right)^2 = \frac{1}{2} \cdot \sum_{p}^{K} \sum_{i}^{N} \left(\sum_{j}^{N} u_{p,j} \cdot w_{i,j} \right) \left(\sum_{j}^{N} u_{p,j} \cdot w_{i,j} \right) =$$

$$= \frac{1}{2} \cdot \sum_{p}^{K} \sum_{i}^{N} u_{p,i} \left(\sum_{j}^{N} u_{p,j} \cdot w_{i,j} \right).$$

(7.4)

At this point it is easy to derive

$$MSE = \frac{1}{2} \cdot \sum_{p}^{K} \sum_{i}^{N} \sum_{j}^{N} u_{p,i} \cdot u_{p,j} \cdot w_{i,j} \qquad (7.5)$$

which is the energy function for an unsupervised ANN (see (7.2)).
Therefore:

$$En = MSE \quad \text{when (target = 0)}. \qquad (7.6)$$

We can thus in principle regard unsupervised ANN learning as a conceptually more economical approach than supervised learning in that it entails doing away with some free parameters, namely, targets. Or, on the other hand, we can make a case for supervised learning, i.e. for the inclusion of the extra free parameters, as a way to focus the learning model upon a more clear-cut task.

Being that it is conceptually (and sometimes practically) easier to build up – and to work with – supervised models (whenever applicable), the literature on unsupervised learning has consequently mainly developed in fields where a supervised approach was out of question, and this has led to a flourishing of studies aimed at an all-round exploration of theoretical issues and possibilities, from developing different models and methods of unsupervised learning, to carefully analyzing their optimality properties under certain conditions (Watkin and Nadal 1994), or providing deep characterizations of their typical error patterns (Liang and Klein 2008). To date, however, there seems to be no reference model that has won an ample consensus in the literature, whereas the menu of possible proposals is surprisingly ample and articulated. In the current situation, setting a standard for unsupervised learning then requires the possible concurrence of two conditions: Conceptual clarity and plausibility of the model, possibly also with a view to the reproduction of actual biological processes, and superior performance.

In this chapter, we provide a new paradigm for unsupervised learning which, on the one side, provides a systematic treatment of the redundancy-based learning mode that is typical of unsupervised learning, and offers a positive model with a possible bio-physical interpretation. On the other side, our approach proves to be systematically outperforming rival supervised learning methodologies for the tasks upon which it has been tested, thereby possibly characterizing suitable forms of unsupervised learning as a *superior* (i.e. both more accurate and more efficient) alternative to supervised learning. According to our approach, supervised learning, rather than enhancing the learning performance due to a clear set-up of the learning task, actually undermines to some extent the learning potential by inefficiently eliminating the redundancy that can only be properly exploited under the unsupervised learning mode. In our view, therefore, unsupervised learning is by no means a 'second best' with respect to supervised learning, the real issue being focusing upon a proper approach to unsupervised learning that is able to unleash its hidden potential.

7.2 GUACAMOLE: A New Paradigm for Competitive Unsupervised Learning – Basic Principles

As it comes to machine learning, a field where supervised ANNs typically obtain good results is intelligent pattern recognition, but when the classification problem becomes particularly complex, they compete with several others statistical tools, such as Support Vector Machines, Bayesian Networks, Classification Trees, KNN, Gaussian Mixtures, Meta-Classifiers, and so on (Duda et al. 2001; Theodoridis and Kotroumbas 2009). All these approaches, however, share a common methodology, that is, they need to be trained on a fraction of the dataset, so that they can rely upon the input-target distinction to calibrate their parameters – in other words, they depend on the partition of the dataset into independent and dependent variables, to generate a specific model to be tested:

$$\text{Data}: \left\{ \mathbf{x_n}, \mathbf{y_n} \right\}_{n=1}^{N}$$

$$\text{Model}: x \rightarrow y = f(x) + \varepsilon.$$

As already remarked, this kind of framing of the problem has no sensible biological counterpart: In real biological cognitive processes, neurons actually create, adapt and delete their connections in an unsupervised way, i.e. treating all the variables on the same logical level. The real issue then becomes building an ANN architecture that is able to carry out sophisticated pattern recognition in an entirely unsupervised way, reasoning in terms of implicit functions rather than of causal models:

$$\text{Data}: \left\{ \mathbf{x_n} \right\}_{n|y_n=K \ K \in [1,2,3,...,C]}^{K}; N = \sum_{k=1}^{C} |\mathbf{x_k}|$$

$$\text{Models}: x_k \rightarrow x'_k = f_k(x_k) + \varepsilon_k.$$

Thus, from the point of view of the first requirement introduced in the previous section, namely biological plausibility, there seem to be reasons to prefer the unsupervised approach over the supervised one. Coming to the second requirement, namely performance, the issue becomes more articulated as there are various dimensions of performance we should take into account, namely, the unsupervised ANN must:

1. Be able to classify input vectors as well as, or better than, classical, supervised algorithms;
2. Once exposed to unfamiliar input vectors, behave as a dynamic memory, i.e., rely upon previously accumulated knowledge to handle new instances in a consistent way – whereas classical supervised algorithms get 'frozen', i.e. they respond unconditionally to the new evidence;
3. Be able to generate on its own new inputs that can be representative of each given class;
4. Be able to simulate the dynamic consequences of its own classifications.

These four sub-requirements into which the notion of performance is split all correspond to natural features of a truly satisfactory pattern classifier. Specifically, (sub) requirement 1 is what we could call *effectiveness*: First of all, of course, patterns have to be properly recognized, at the highest possible standard against competitor systems. Requirement 2 may be termed *flexibility*: The system is able to generalize its knowledge to handle unprecedented stimuli in an appropriate way. Requirement 3 is what one means for *specificity*: the system understands the typology it has constructed, and is able to replicate the characteristic aspects of each of its classes. Finally, requirement 4 calls for *insightfulness*: the system is able to figure out how the constructed categories will tend to adapt to changing circumstances.

In this paper, we take this challenge and propose an unsupervised ANN which meets in principle all of the four requirements above. It is called GUACAMOLE **(General Unsupervised Adaptive Classification Algorithm for Modular Orga-nization of Learning Evolution)**. The basic principle behind GUACAMOLE is very intuitive: It feeds upon a collection of Auto-Associative ANNs (Buscema and Sacco 2010), each of which is trained to be a 'specialist' of a given class of the recognition task. Each ANN focuses upon its own assigned class, without even communicating with the others, and learns in an unsupervised way without any target. 'Recognizing' a pattern as belonging to a certain class then amounts to designing a competitive scheme that at every given round of testing/prediction chooses as the winner the 'expert ANN' for the appropriate category: The key point is thus how to design such an allocation scheme, and this is what qualifies GUA-CAMOLE irrespectively of the specific Auto-Associative ANNs which are employed as 'experts'.

More specifically, the whole procedure may be described as follows. Given the recognition problem with N given classes, GUACAMOLE calls for N Auto-Associative ANNs, one for each class. The available database is split in a random way into a Training and a Testing sub-database, respectively. At the beginning of the Training stage, the Training sub-database is further split into N sub-sub-databases, one for each class of the recognition task, and the training records are assigned accordingly to the appropriate class (database). At this point, Training takes place, and each ANN learns about its own assigned class, thus becoming the system's 'expert' for that class. It is important to stress that each ANN only sees the training records pertaining to its assigned class, and knows nothing about other classes nor how the other ANNs are faring with their training. At the end of the Training phase, the weights for each ANN are 'frozen'.

In the Testing phase, then, each ANN keeps its weights fixed and does not learn further from experience. Each record from the Testing sub-database is then submitted to *each* of the ANNs, and the corresponding errors are computed at fixed weights. At this point, the competitive allocation rule is put to work. The rule simply prescribes that the ANN whose Testing error is closer to the average error incurred in the Training phase is the winning one, i.e., the test record is attributed to its class. Intuitively, we could say that the winning ANN is the one that results in being less 'surprised' by the appearance of that specific test record, i.e. the one that finds it more 'typical' given its class-specific training background. It is apparent here how this particular approach to unsupervised learning relates in a direct way to the knowledge-generating value of redundant information, and on the construction of reference types for each taxonomic class.

7.3   GUACAMOLE: Formal Treatment

We are now ready to present the GUACAMOLE architecture in more formal terms. In the first place, we have to define the particular kind of Auto-Associative ANNs that are used to implement GUACAMOLE in this specific instance. The choice has been to work with New Recirculation ANNs (NRC henceforth), i.e., an enhanced version of Hinton's Recirculation ANN (Hinton and McClelland 1988) developed in Buscema (1998). The peculiarity of the NRC is that it is an unsupervised Auto-Associative ANN which learns through a comparison between its own expectations about the outcome of its own learning, and the actual outcome, i.e., by reasoning about, and interpreting, its own learning process. What NRC actually does is to re-circulate, literally, the output that has been produced at the end of a processing cycle as a new input for the next processing cycle, and keeping on like this until the process converges, i.e. until the process reaches a stationary state. In other words, the arrival of a new input causes NRC to 'ruminate' the new piece of information until it has been digested by the system: In this unsupervised process, it is as if the ANN takes as its *provisional* target the actual output of the previous processing cycle, so that the process ends when the 'target' is reached, i.e., when a new processing cycle does not cause further modifications of the weights.

In formal terms, the equations defining NRC are the following:

Real Output calculation:

$$a_o^R = f\left(\sum_i w_{o,i}^{[n]} \cdot a_i^R + \theta_o^{[n]}\right) = f\left(Net_o^R\right) = \frac{1}{1 + Exp\left(-Net_o^R\right)}; \qquad (7.7)$$

Imaginary Input calculation:

$$Net_i^I = \sum_o w_{i,o}^{[n]} \cdot a_o^R + \theta_i^{[n]}; \qquad (7.8a)$$

$$a_i^I = r \cdot a_i^R + (1 - r) \cdot \frac{1}{1 + e^{-Net_i^I}}; \ 0 < r < 1 \qquad (7.8b)$$

Imaginary Output calculation:

$$Net_o^I = \sum_i w_{o,i}^{[n]} \cdot a_i^I + \theta_o^{[n]}; \qquad (7.9a)$$

$$a_o^I = r \cdot a_o^R + (1 - r) \cdot \frac{1}{1 + Exp\left(-Net_o^I\right)}; \qquad (7.9b)$$

Weights error calculation:

$$\Delta w_{i,o} = Rate \cdot a_o^R \cdot \left(a_i^R - a_i^I\right); \ 0 < Rate \leq 1 \qquad (7.10a)$$

$$\Delta w_{o,i} = Rate \cdot a_i^I \cdot \left(a_o^R - a_o^I\right); \qquad (7.10b)$$

Biases error calculation:

$$\Delta\theta_o = Rate \cdot \left(a_o^R - a_o^I\right); \qquad (7.11a)$$

$$\Delta\theta_i = Rate \cdot \left(a_i^R - a_i^I\right); \qquad (7.11b)$$

Weights updating:

$$w_{i,o}^{[n+1]} = w_{i,o}^{[n]} + \Delta w_{i,o}; \qquad (7.12a)$$

$$w_{o,i}^{[n+1]} = w_{o,i}^{[n]} + \Delta w_{o,i}; \qquad (7.12b)$$

Biases updating:

$$\theta_o^{[n+1]} = \theta_o^{[n]} + \Delta\theta_o; \qquad (7.13a)$$

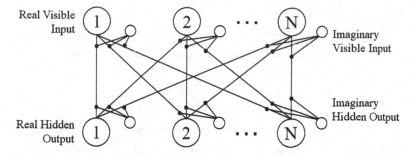

Fig. 7.1 Topology of the new recirculation ANN

$$\theta_i^{[n+1]} = \theta_i^{[n]} + \Delta\theta_i. \qquad (7.13b)$$

where r is a projection coefficient and *Rate* is the learning coefficient. To understand the equations, we must introduce more detail about the actual structure of NRC, as in Fig. 7.1 below.

The structure of NRC distinguishes between two different typologies of input and output: The *Real* and the *Imaginary* (hypothetical) ones. Real values are denoted by an R superscript, whereas imaginary ones by an I superscript. In turn, Input values are denoted by an i subscript, whereas Output values by an o subscript. Given the nature of NRC, we consider a generic signal a, without distinguishing in the main notational corpus between input and output values (which are instead identified by subscripts), as the same signal plays either the input or the output roles at different points of the learning cycle. Thus, accordingly, by a_i^I and a_o^I we mean the imaginary input and output signals, respectively, and likewise for a_i^R and a_o^R, the real input and output signals. Moreover, by w_{oi} we mean the (down-streaming) weights from the input to the output layer, and by w_{io} the (up-streaming, recirculation) weights from the output to the input layer. Finally, the θs are the biases.

Thus, (7.7) tells us that NRC produces its real output by a classical sigmoid filtering of its net real input. Real output, in turn, becomes the net input for the imaginary part of the ANN (7.8a); in particular, NRC's *imaginary* input is a weighted average (according to the projection coefficient r) of the *real* input and of the net real output at the *previous* stage (7.8b): That is to say, NRC directs its learning process by reflecting on the outcomes of its own learning process at previous stages and by constructing a new, idiosyncratic evidence base which is the blending, according to a given relative weight, of the external evidence and of its own past rumination of the same evidence. The imaginary inputs thus generated become the basis for an imaginary output, which is obtained by the sigmoid filtering of the net imaginary input (7.9b): That is to say, NRC elaborates an 'opinion' of its own by processing in the familiar way its own idiosyncratic evidence base, which becomes the hypothetic benchmark against which to compare the actual output. The way NRC learns, therefore, is by comparison between the outcomes of the real and of the imaginary information processing, whereas the latter directly takes into account the NRC's own architecture, i.e., is the product of self-reflection.

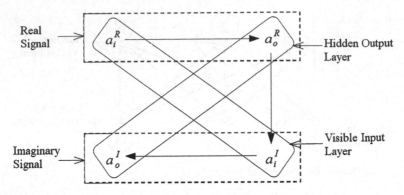

Fig. 7.2 Dynamics of signal transfer in NRC

Up-streaming and down-streaming weights, accordingly, are revised according to the distance between the real and imaginary outputs and inputs, respectively – that is to say, the distance between the real and the imaginary magnitude of the signal is interpreted by the system as the 'error' to be optimized (7.10b). Biases are revised according to the same logic (7.11b). Finally, (7.12a and 7.12b) and (7.13a and 7.13b) describe weights and biases updating, respectively.

This brief discussion should thus elucidate in what sense NRC uses the real magnitudes as a 'moving target' to adjust its conjectural (imaginary) structure: The ANN keeps on adjusting itself, i.e. on self-redesigning its own 'imagination', until the latter turns out to fit naturally the actual observed evidence. Once the observed evidence turns out to be no longer 'surprising' on the basis of the ANN's 'imagination', then the learning process cools down and NRC 'loses interest' in further processing of the same data. A synthetic picture of the actual structure of the informational flow for the NRC is provided in Fig. 7.2 below.

Choosing NRCs as 'local experts' for GUACAMOLE's optimal classification strategy means that, in the Training phase, each NRC develops an internal representation of the structure of the particular class to which it has been assigned, and by doing so, indirectly constructs a characterization of the typical features that identify that specific class. Once the NRCs are exposed to the Testing records during the Testing phase, they tend therefore to be more or less 'surprised' about the characteristics of the record they are being shown depending on the extent to which such characteristics reflect the typicality of their class. This is the simple principle on which the formal architecture of GUACAMOLE is based:

Legend :

$a^R_{i,j}$ = i-th input of the j-th pattern;
$a^I_{i,j}$ = i-th output of the j-th pattern;
Q = Number of Input and Output units;
M_p = Number of Patterns of the pth ANN;
ANN_p = pth ANN; $p \in N$;
N = Number of the ANNs and consequently number of the Classes.

$$TrainError_p = \frac{\sum_{j}^{M_p} \sum_{i}^{Q} \left(a_{i,j}^R - a_{i,j}^I\right)^2}{M_p}; \qquad (7.14)$$

$$TestError_p = \sum_{i}^{Q} \left(a_i^R - a_i^I\right)^2; \qquad (7.15)$$

$$En_p = \left(TrainError_p - TestError_p\right)^2; \qquad (7.16)$$

$$ANN_{winner} = \arg \underset{p}{Min} \{En_p\}. \qquad (7.17)$$

GUACAMOLE then defines the Training error of each NRC in terms of the discrepancy between its real and imaginary inputs for the Training database, and the Testing error in terms of the discrepancy between its (fixed weights) real and imaginary inputs for that specific Testing record. The winning expert is the one for which the discrepancy between the above defined Training and Resting errors is minimal, that is to say, the NRC for which the given Testing record elicits an imaginary response (on the basis of the *fixed* weights determined by the shaping up of the NRC's imaginary system during the Training phase) which best fits its pre-existent representation of the levels and modes of variability of 'its' class. This implies, by the way, that the winning NRC need not be the one that commits the smallest Testing error – the winning NRC is thus not the most accurate, but the one for which the testing error is as close as possible to the magnitude of its Training error, i.e. the NRC whose performance in the Testing phase better resonates with the performance in the Training phase given the specific structural complexity of the class about which it has been assigned to learn. A representation of the GUACAMOLE architecture during the Training and Testing phases, respectively, is provided in Figs. 7.3 and 7.4.

7.4 Benchmarking GUACAMOLE in Pattern Recognition Tasks

After having presented the structure of GUACAMOLE, we are finally in the position to evaluate its performance against a battery of the best available supervised and unsupervised systems in a pattern recognition task, thereby meeting one of the two requirements put forward in the introductory section. We have chosen in particular the following competitor systems: the Parzen (1962) classifier (PARZENC); the BackPropagation ANN (Chauvin and Rumelhart 1995) (BackProp); the Extended Delta Bar Delta ANN (Minai and Williams 1990) (EDBD); the Learning Vector Quantization (Kohonen 2001) (LVQ); the Support

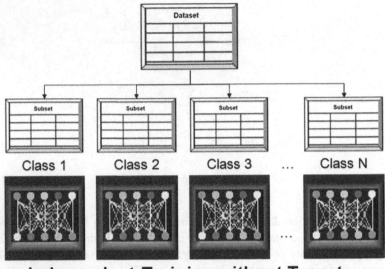

Fig. 7.3 GUACAMOLE training

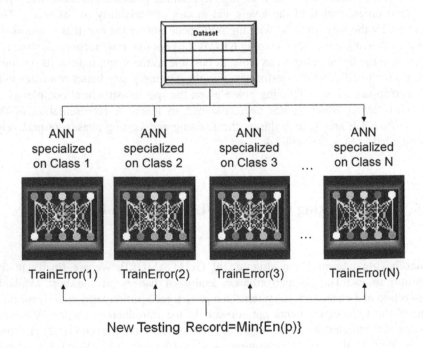

Fig. 7.4 GUACAMOLE testing

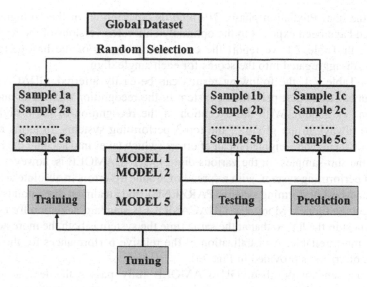

Fig. 7.5 Validation protocol

Vector Machine (Cortes and Vapnik 1995) (SVM); the Modular Neural Networks (Jordan and Jacobs 1992) (MNN); the Linear Discriminant Classifier (McLachlan 2004) (LDC); the Quadratic Discriminant Classifier (McLachlan 2004) (QDC); the K-Nearest Neighbor (Bremner et al. 2005), and the Naïve Bayes Classifier (Zhang 2004) (NAIVEBC). Overall, this battery of techniques represents a fair covering of the most diffused and credited machine learning tools for pattern recognition from a variety of conceptual backgrounds and analytical approaches, sampling both inside and outside the ANN field.

The first specific task that has been chosen for the benchmark is the Digit Database for the recognition of hand-written numeric characters to be found in the UCI Repository at the address http://archive.ics.uci.edu/ml/. It is a database of 1,594 hand-written digits by different subjects in different situations, which have been codified in a 256 bit streak, corresponding to a 16×16 grid. The task is to assign each grid to the corresponding digit, for the ten 0–9 digits.

The validation protocol chosen is the familiar one with three phases: Training, Testing, and Prediction. The database is randomly split into the Training, Testing and Prediction sets, respectively, and is formatted according to the needs of the specific system being employed. Systems are optimally calibrated during the Training phase, and then the performance of the calibrations is measured under the Testing phase without further learning. Wherever necessary, a Tuning sub-phase is carried out. Finally, all rival systems are exposed to the Prediction phase, where entirely new records are shown and their one-shot comparative performance is measured. Five independent blind runs have been conducted for each system, and the results reported represent an average of the outcomes of the five runs. A synthetic representation of the validation protocol can be found in Fig. 7.5.

For the final Prediction phase, 319 records covering all of the 10 digits to be classified have been exposed to the optimally calibrated versions of the respective systems. In Table 7.1 we report the comparative scores of the best-performing systems, disaggregated into the scores for each single digit:

From Table 7.1, the following results can be easily inferred. GUACAMOLE turns out to be the best performing system in the recognition of 5 digits out of 10 (with one ex-aequo). With the exception of the recognition of the 0 digit, it is however always in the group of the top-3 performing systems. In terms of both arithmetic mean and weighted mean of errors (which takes into account the relative size of the sub-samples for the various digits), GUACAMOLE is, however, by far the best performing system, with a spread of around 10 errors in absolute size from the second-best performing system (PARZENC), and an almost three-points spread in percentage figures. Moreover, GUACAMOLE's standard deviation for errors is the smallest in the lot, so that at the same time the system is both the more accurate and the more reliable. A visualization of the relative performances for the simple average of errors is provided in Fig. 7.6.

In this benchmark, then, GUACAMOLE fully passes the test of relative performances with respect to competing supervised and unsupervised systems. Moreover, the performance gap is such that the benchmark seems to confirm the idea that a carefully designed unsupervised system may be inherently superior to alternative supervised systems once the knowledge potential of redundancy is properly channeled, and not, as it has been often argued, as a second-best alternative to be used when the structure of the problem does not allow for supervised learning.

The second benchmark test we carry out has to do with a more realistic situation, namely, the recognition of patients of Amyotrophic Lateral Sclerosis (ALS), a complex, highly multi-factorial disease which seems to derive from a subtle interaction between environmental and genetic susceptibility factors. The database is made of 61 polymorphisms within 35 genes, sampled in 54 sporadic ALS patients and 208 controls. A preliminary study of ANN-based pattern recognition on this database has been conducted in Penco et al. (2008). Ability to successfully detect ALS patients from genetic background then provides a clear case for complex, multi-factorial genetic background as a major and only partially understood determinant of the disease. The results in Penco et al. (2008) already provided sound evidence of this sort, with the top scoring system reaching an overall average success in prediction of around 96 %. Here we use this same database as a second benchmark for our battery of competing systems already tested out with the Digit database.

In this second experiment, we have adopted an even more articulated validation protocol, in which not only, as before, five independent runs for each system are carried out, but each system is in turn Trained and Tested twice in order to rule out the possibility of selecting sub-optimal calibrations for a given system. The new validation protocol is reported in Fig. 7.7.

As before, we report here in Table 7.2 the scores obtained by the best performing systems, as averages for the 5 × 2 independent runs, for a Prediction test database of 265 records (58 cases vs. 207 controls).

Table 7.1 Comparative scores of the best performing systems for the digit database
319 patterns of Handwritten Digit : 256 Input × 10 Classes

Mean of five independent blind test	Zero (%)	One (%)	Two (%)	Three (%)	Four (%)	Five (%)	Six (%)	Seven (%)	Eight (%)	Nine (%)	Arithmetic mean (%)	Weighted mean (%)	Errors	Errors St Dev
GUACAMOLE	96.29	94.47	93.09	94.98	95.68	94.35	97.50	96.81	92.90	87.94	94.40	94.41	17.8	4.76
PARZEN	98.13	98.16	93.13	94.98	88.86	95.59	95.65	91.80	81.94	76.59	91.48	91.53	27	5.87
EDBD	95.65	90.80	89.33	90.60	89.49	93.06	96.27	89.90	86.45	86.07	90.76	90.77	29.4	5.03
LVQ	98.13	92.61	91.25	89.33	87.05	91.80	93.16	87.32	89.03	81.01	90.07	90.08	31.6	6.58
SVM	90.68	86.44	86.88	88.09	90.78	89.96	90.66	92.36	97.42	85.36	89.86	89.83	32.4	8.76
MNN	94.38	87.67	87.44	88.09	88.84	92.44	96.25	84.23	87.10	85.38	89.18	89.20	34.4	5.55
LDC	95.00	88.24	88.06	90.62	85.80	88.69	94.41	84.11	78.71	85.48	87.91	87.95	38.4	10.97
KNN	98.75	98.79	93.75	94.92	90.74	94.96	95.00	74.05	66.45	56.98	86.44	86.57	42.8	6.38
NAIVEBC	94.41	77.84	82.42	88.73	79.54	85.52	88.81	92.36	82.58	79.05	85.13	85.12	47.4	9.81

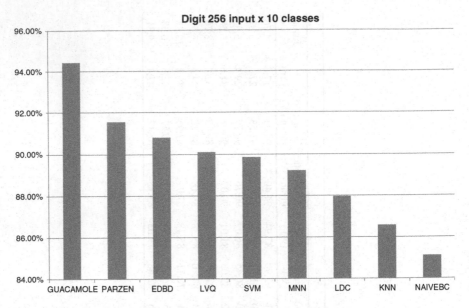

Fig. 7.6 Relative performance of the competing systems, digit database

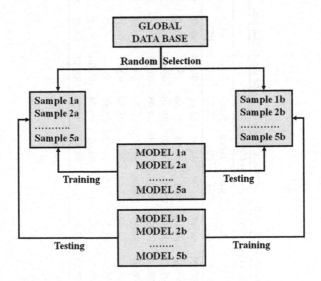

Fig. 7.7 5 × 2 validation protocol

As shown by the results, GUACAMOLE is neither the best system in detecting actual patients nor in detecting controls, but overall it turns out to be by far the best performing system both in terms of accuracy and reliability: The two rival systems whose standard deviation is smaller than GUACAMOLE are the two worst performing ones, and thus in this case a low deviation means that they are stuck

Table 7.2 Comparative scores of the best performing systems for the ALS database

Amyotrophic Lateral Sclerosis 61 input × 2 classes

265 Records: 58 Cases vs. 207 Controls

5 × 2CV	Cases (%)	Controls (%)	A.Mean Acc (%)	W.Mean Acc (%)	Errors	St Dev
GUACAMOLE	93.79	99.61	96.70	98.34	2.2	1.32
SVM	87.59	98.07	92.83	95.77	5.6	1.96
Back Prop	80.35	80.20	80.27	80.23	26.2	8.02
LDC	84.10	70.85	77.48	75.53	33.1	4.48
NAIVEBC	96.90	50.00	73.45	60.25	52.7	53.67
KNN	100.00	12.55	56.28	31.69	90.5	3.03
Quadratic DC	13.47	99.24	56.36	44.68	91.5	0.53
PARZENC	100.00	0.10	50.05	21.96	103.4	0.7

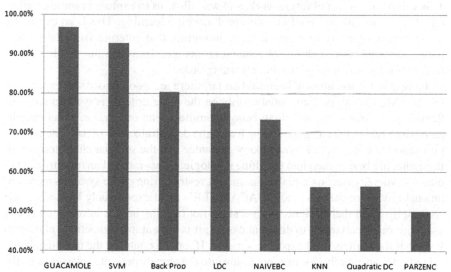

Fig. 7.8 Relative performance of the competing systems, ALS database

away from an acceptable level of performance. The second-best system in this benchmark, namely SVM, is such that even the 1-standard deviation score above weighted mean remains below the weighted mean of GUACAMOLE: Once again, thus, there is a performance that is on a different order of magnitude than that of the best rival system. The comparative performance scores are visualized in Fig. 7.8.

The superior performance of GUACAMOLE in entirely different recognition tasks, in which by the way the performance of rival systems varies wildly (for instance, the second-best performing system for the Digit benchmark, PARZENC, is the worst performing one in the ALS benchmark), seems to suggest that indeed this type of unsupervised system may set a new standard, opening a new view on the potential of unsupervised learning systems as the natural option even in context where traditionally supervised systems are electively chosen and considered to be the reference option.

7.5 Discussion

In this final section, we briefly discuss some of the implications of our results for the requirements set forth in the introductory section of the paper and for future research. We have already seen that, speaking of performance, it is indeed the case that our unsupervised system outperforms the rivals in terms of both predictive accuracy and reliability. The effectiveness sub-criterion is then strongly met by GUACAMOLE. At this point, it will be important to test the system in terms of the remaining performance criteria, namely, flexibility, specificity, and insightfulness. This will be the object of future work, and will allow us to explore in some detail so far often overseen aspects of unsupervised machine learning. This is an especially interesting perspective in view of the main finding that emerges from the present paper, namely, the possibility of looking at unsupervised learning as the *natural* framework for *any* kind of machine learning task.

From what it has already been said so far, there are good reasons to expect that GUACAMOLE can perform satisfactorily on the other criteria as well. In terms of flexibility, insofar as the new item being submitted is an outlier, i.e. an anomalous member of one of the categories being learnt, the decentralization of the system into a population of local experts would likely guarantee that the singular characteristics of the outlier are best framed into the affine typological class rather than misattributed to others – whereby causing a *global*, inefficient restructuring of the system's representational model for the task – as GUACAMOLE's local experts only learn about the characteristics of their assigned class and are not messing up this information in any way with other pertaining to different classes; it is only at the competitive allocation level that the various local expertise's merge. If, on the contrary, the new item being submitted is an entirely new one, not fitting into the already presented classes (e.g., the number 10, made of two digits rather than one, for the Digit database), again the system can adapt reasoning by analogy, i.e. by assigning the new number to the class that corresponds to the local expert that is 'surprised' the least by the new occurrence – and given the nature of the typology constructed by the local expert, it is likely that the attribution will occur on the basis of structural rather than contingent characteristics, and that therefore it will display a relative stability across cases.

Likewise, and by the same token, elicitation of new, unobserved representative examples of a given class is a relatively easy task for the local expert of GUACA-MOLE, unlike what would happen for other methods which do not work upon a separable process of typology-building for the various classes. And finally, given the self-reflecting nature of the Auto-Associative ANNs employed in GUACA-MOLE (and most notably of the NRC used in the implementation carried out in this paper), this system is naturally qualified to explore the dynamic implications of its own categorizations. All these intuitions, however, will be rigorously tested in future work.

Finally, it is interesting to discuss how the GUACAMOLE architecture responds to the other requirement put forth, namely, its conceptual clarity and biological plausibility. The idea that pattern recognition may be managed through the development of local specialists which are independently trained for the purpose is fascinating and not devoid of biological plausibility. The available preliminary evidence seems to suggest that distributed neural systems play a key role in classification tasks, and that distinct aspects of the task tend to be mediated by distinct neural representations (e.g. Haxby et al. 2000). Although at the moment it is not possible to conjecture any specific counterpart in the biological organization of human cognition, the proposed one is nevertheless a clear-cut logical scheme which may be the object of finalized thinking and analysis, unlike what generally happens for most of the competing machine learning systems which do not admit an analogous biological scheming – let alone an accurate translation into biological terms.

For these reasons, we look forward to further work to check whether GUACA-MOLE may be seen as a first instance of a new paradigm for unsupervised learning – and for certain dimensions of machine learning more generally.

References

Bailey TL, Elkan C (1995) Unsupervised learning of multiple motifs in biopolymers using EM. Mach Learn 21:51–80

Barlow HB (1989) Unsupervised learning. Neural Comput 1:295–311

Bremner D, Demaine E, Ericson J, Iacono J, Langerman S, Morin P, Toussant G (2004) Output-sensitive algorithms for computing nearest-neighbor decision boundaries. Discrete Comput Geometry 33:593–604

Buscema M (1998) Recirculation neural networks. Subst Use Misuse 33:383–388

Buscema M, Sacco PL (2010) Auto-contractive maps, the H function, and the maximally regular graph: a new methodology for data minino. In: Capecchi V, Buscema M, Cantucci P, D'Amore B (eds) Applications of mathematics in models, artificial neural networks and arts, Mathematics and Society. Springer, Dordrecht, pp 227–275

Carneiro G, Chan AB, Moreno PJ, Vasconcelos N (2007) Supervised learning of semantic classes for image annotation and retrieval. IEEE Trans Pattern Anal Mach Intell 29:394–410

Chapelle O, Schölkopf B, Zien A (eds) (2006) Semi-supervised learning. MIT Press, Cambridge, MA

Chauvin Y, Rumelhart DE (1995) Back propagation: theory, architecture, and applications. Lawrence Erlbaum Associates, Hillsdale

Cortes C, Vapnik V (1995) Support-vector networks. Mach Learn 20:273–297

Damper RI, French RLB, Scutt TW (2000) ARBIB: an autonomous robot based on inspiration from biology. Robot Auton Syst 31:247–274

Duda RO, Hart PE, Stork DG (2001) Pattern classification, 2nd edn. Wiley, New York

El-Yaniv R, Souroujon O (2001) Iterative double clustering for unsupervised and semi-supervised learning. In: Advances in neural information processing systems. MIT Press, Cambridge, MA, pp 1025–1032

Gureckis TM, Love BC (2003) Towards a unified account of supervised and unsupervised category learning. J Exp Theor Artif Intell 15:1–24

Hansen LK, Larsen J (1996) Unsupervised learning and generalization. In: Proceedings of the IEEE international conference on neural networks, Washington, DC

Haxby JV, Hoffman EA, Gobbini MI (2000) The distributed human neural system for face perception. Trends Cogn Sci 4:223–233

Hinton GE, McClelland JM (1988) Learning representation by recirculation. In: Anderson DZ (ed) Neural information processing systems. American Institute of Physics, New York, pp 358–366

Japkowicz N (2001) Supervised versus unsupervised binary-learning by feedforward neural networks. Mach Learn 42:97–122

Jordan MI, Jacobs RA (1992) Hierarchies of adaptive experts. Adv Neural Inf Process Syst 4:985–992

Kim YS, Street WN, Menczer F (2002) Evolutionary model selection in unsupervised learning. Intell Data Anal 6:531–556

Ko Y, Seo J (2000) Automatic text categorization by unsupervised learning. In: Proceedings of the 18th conference on computational linguistics, vol 1, Saarbrücken

Kohonen T (2001) Self-organizing maps. Springer, Berlin

Körding KP, König P (2001) Supervised and unsupervised learning with two sites of synaptic integration. J Comput Neurosci 11:207–215

Lange T, Braun ML, Roth V, Buhmann JM (2002) Stability-based model selection. In: Advances in neural information processing systems. MIT Press, Cambridge, MA, pp 617–624

Liang P, Klein D (2008) Analyzing the errors of unsupervised learning. In: Proceedings of ACL-08: HLT, Columbus, pp 879–887

Love BC (2002) Comparing supervised and unsupervised category learning. Psychon Bull Rev 9:829–835

Malakooti B, Raman V (2000) Clustering and selection of multiple criteria alternatives under unsupervised and supervised neural networks. J Intell Manuf 11:435–451

McLachlan GJ (2004) Discriminant analysis and statistical pattern recognition. Wiley, New York

Minai AA, Williams RD (1990) Acceleration of back-propagation through learning rate and momentum adaptation. In: Proceedings of the IEEE/INNS international joint conference on neural networks, vol 1, Washington DC, pp 676–679

Nadal JP, Parga N (1994) Duality between learning machines: a bridge between supervised and unsupervised learning. Neural Comput 6:491–508

Parzen E (1962) An estimation of a probability density function and mode. Ann Math Stat 33:1065–1076

Penco S, Buscema M, Patrosso MC, Marocchi A, Grossi E (2008) New application of intelligent agents in sporadic amyotrophic lateral sclerosis identifies unexpected genetic background. BMC Bioinformatics 9:254–266

Sutton RS, Barto AG (1998) Reinforcement learning: an introduction. MIT Press, Cambridge, MA

Theodoridis S, Kotroumbas K (2009) Pattern recognition, 4th edn. Academic, New York

Watkin TLH, Nadal JP (1994) Optimal unsupervised learning. J Phys A 27:1899–1915

Xu L (1994) Multisets modelling learning: an unified theory for supervised and unsupervised learning. In: Proceedings of the IEEE international conference on neural networks, vol I, Orlando, pp 315–320

Yarowsky D (1995) Unsupervised word sense disambiguation rivaling supervised methods. In: Proceedings of the 33rd annual meeting of the association for computational linguistics, Cambridge, MA

Zhang H (2004) The optimality of naïve Bayes. In: Barr V, Markov Z (eds) Proceedings of the seventeenth international Florida artificial intelligence research society conference. AAAI Press, Miami Beach

Zhao Z, Liu H (2007) Spectral feature selection for supervised and unsupervised learning. In: Proceedings of the 24th international conference on machine learning, Corvallis

Zhu X, Goldberg A (2009) Introduction to semi-supervised learning. Synthesis lectures on artificial intelligence and machine learning, vol 3. Morgan & Claypool Publishers, San Rafael, pp 1–130

Chapter 8
Spatiotemporal Mining: A Systematic Approach to Discrete Diffusion Models for Time and Space Extrapolation

Massimo Buscema, Pier Luigi Sacco, Enzo Grossi, and Weldon A. Lodwick

8.1 Introduction: Diffusion as an Ubiquitous Concept

Diffusion may be synthetically described as the phenomenon of spatiotemporal propagation of a certain variable across a medium, where the modes of propagation crucially depend on the characteristics of the medium and the entities of interest. Instances of propagation abound in virtually every discipline, and thus there is a rich, vast and heterogeneous literature dealing with it. Consequently, diffusion may safely be deemed to be a ubiquitous concept (Buchanan 2001). A consequence of this ubiquity is that diffusion has been tackled analytically by scholars belonging to very different disciplines, using a diverse range of techniques. In some cases the medium has a social nature such as in the case of epidemics or in the propagation of innovations and attitudes, whereas in others it has a physical nature, like in the propagation of a fluid through a porous medium or in the propagation of earthquakes in the underlying crust.

In this study we present a general approach to the study of diffusion phenomena based on a relatively simple though rather general framework, and on an entirely new mathematical treatment that allows for accurate description/reproduction, understanding, and prediction of the outcomes of diffusion phenomena. Our approach is the result of the combination of several modules that have been

M. Buscema (✉)
Semeion Research Center of Sciences of Communication, Via Sersale 117, Rome, Italy

Department of Mathematical and Statistical Sciences, CCMB, University of Colorado, Denver, Colorado, USA
e-mail: m.buscema@semeion.it

P.L. Sacco • E. Grossi
Semeion Research Center of Sciences of Communication, Via Sersale 117, Rome, Italy

W.A. Lodwick
Center for Computational and Mathematical Biology, Department of Mathematics, University of Colorado, Denver, CO, USA

W.J. Tastle (ed.), *Data Mining Applications Using Artificial Adaptive Systems*,   231
DOI 10.1007/978-1-4614-4223-3_8, © Springer Science+Business Media New York 2013

developed recently at Semeion Research Center, and in particular what we call the Target Diffusion Model (TDM)[1] and Twisting Algorithm based on Twisting Theory (TWT).[2] The aim of this chapter is to present our general approach to the modeling of diffusion phenomena and its implementation through the TDM + TWT methodology. The approach is subsequently tested for predictive accuracy using as a benchmark epidemic data from the Dengue fever crisis that occurred in the state of São Paulo, Brazil, in 2001 (Ferreira 2005, 2010).

More specifically, we may think of diffusion in terms of the following general logical scheme. Consider M fixed entities in two-dimensional space (the entity set), and a sequence of discrete time periods $n = 1, \ldots, N$ where to each of the entities one can assign a vector of state variables which describe the overall state of the entity at each given time with respect to the phenomenon under study. For example, the entities could be specific population centers and the vector of state variables could describe the number of infected and non-infected residents at each given time, respectively. Or alternatively, to recall an issue that has become very popular in the study of the diffusion of innovations, the entities could be physicians and the state variables could be the medical treatments that they prescribe to their patients for given diseases at each given time (Becker 1970). The reason we refer to an entities set is that our measurement capacity is limited, and therefore we are generally able to describe the diffusion of the variables across the medium only at carefully chosen sample points. The underlying assumption is, of course, that there are spatiotemporal cause/effect relationships that are captured by our models from the data. Our approach is to let the data determine what these relations are rather than imposing *a-priori* relationships as would occur in partial differential equation diffusion models. To be sure, there is an underlying architecture to how the data are analyzed to extract the dynamics. In order to understand the actual mechanics of diffusion, we must address three basic issues:

(a) *The reconstruction of the global causation process*: what are the entities whose state at time n influences any given entity at time $n + 1$? And with what strength? That is to say, what is the global causation process that drives the diffusion of the quantity under study through the medium? We will be able to address this issue by means of the TDM.

(b) *The reconstruction of the effects on unmapped entities*: what can be said of the effects of the diffusion process for entities (e.g. at points) which are not included in the original entities set but belong to the same diffusion space.

---

[1] Target Diffusion Model (T.D.M.) is a USA Patent pending#13/070,854 (24 March 2011), inventor: Massimo Buscema, m.buscema@semeion.it; owners Semeion Research Center of Sciences of Communication, via Sersale 117, Rome, 00128, Italy and CSI Research & Environment, via CesarePavese, 305, Rome, Italy.

[2] Twisting Theory is a USA Patent pending #12/969,887 (16 Dec 2010), inventor: Massimo Buscema, m.buscema@semeion.it; owners Semeion Research Center of Sciences of Communication, via Sersale 117, Rome, 00128, Italy and CSI Research & Environment, via CesarePavese, 305, Rome, Italy.

For example, what are the diffusion patterns at points that are not part of the original population centers data? Using the outputs of TDM, we will apply TWT to address this issue.

(c) *The step-by-step predictability of the process*: what are the invariants of the process, that is to say, is it possible to predict which entities at n will influence what entities at $n + 1$ in a blind (i.e., unconditional) way? Using the outputs of TDM, we will make use of a specific supervised artificial neural network to address this issue.

Being able to successfully tackle issues a–c above basically amounts to being able to decipher and reproduce the inner mechanics (also called the behavior system) of the diffusion process. As our approach is not based upon any specific disciplinary perspective, we believe that if such goal is reached, the corresponding methodology must be successfully applied to any instance of diffusion phenomenon, whatever the disciplinary context in which it takes place. Therefore, in this paper we are up at developing an all-purpose approach to the study of diffusion. To emphasize, our philosophy is thus one in which we impose no *a-priori* structure such as is true in partial differential equation models on the dynamical system. We seek to extract this structure from the available data. We will, however, impose an architecture to the way we extract relationships, but we do not predefine the relationships as is true of many other approaches such as partial differential equations.

8.2 The Target Diffusion Model (TDM)

As explained in the introductory section, we consider here M entities in the two-dimensional plane, identified by their Cartesian coordinates (X,Y), observed at a sequence of times $n = 1,\ldots,N$. To each entity is assigned a quantity that varies through time, and that is described by an $M \times N$ matrix Q. We can resume the above in the following table (Table 8.1):

To reconstruct the causation process that drives diffusion, we have to build a model of the reciprocal influences among entities through time. The basic elements that we have to consider are: The spatial coordinates of the entities (Space, S); the quantities assigned to each entity (Quantity, Q), at each time step (Time, T), so that we can write the causation model in its general form as

$$M = \Psi(S, T, Q).$$

We can accordingly define a local model $M([n, n + 1])$ in Markovian terms, to express the likelihood of relationship between any entity at time n and any entity at time $n + 1$, in the following way:

$$M_{i,j}^{[n,n+1]} = \Psi\left(x_i, y_i, q_i^{[n]}, q_i^{[n+1]}, x_j, y_j, q_j^{[n]}, q_j^{[n+1]}\right).$$

Table 8.1 Entities, their coordinates across the medium and the time evolution of the quantity

Entities	X	Y	Time (1)	Time (2)	Time (3)	Time (...)	Time (N)
Entity (1)	Ex(1)	Ey(1)	q(1,1)	q(1,2)	q(1,3)	q(1,...)	q(1,N)
Entity (2)	Ex(2)	Ey(2)	q(2,1)	q(2,2)	q(2,3)	q(2,...)	q(2,N)
Entity (3)	Ex(3)	Ey(3)	q(3,1)	q(3,2)	q(3,3)	q(3,...)	q(3,N)
Entity (4)	Ex(4)	Ey(4)	q(4,1)	q(4,2)	q(4,3)	q(4,...)	q(4,N)
Entity (5)	Ex(5)	Ey(5)	q(5,1)	q(5,2)	q(5,3)	q(5,...)	q(5,N)
Entity (6)	Ex(6)	Ey(6)	q(6,1)	q(6,2)	q(6,3)	q(6,...)	q(6,N)
Entity (7)	Ex(7)	Ey(7)	q(7,1)	q(7,2)	q(7,3)	q(7,...)	q(7,N)
Entity (8)	Ex(8)	Ey(8)	q(8,1)	q(8,2)	q(8,3)	q(8,...)	q(8,N)
Entity (...)	Ex(...)	Ey(...)	q(...,1)	q(...,2)	q(...,3)	q(...,...)	q(...,N)
Entity (M)	Ex(M)	Ey(M)	q(M,1)	q(M,2)	q(M,3)	q(M,...)	q(M,N)

The TDM translates this general scheme into a specific model of diffusion dynamics, which can be thought of as a 'competitive gravity' model, where the strength of the link between two entities is measured by the direct product of their quantities, (exponentially) weighted by their relative distance, and adjusted by their relative increments. Among the various strengths of association between one entity and all of the others, the stronger one 'wins' and causes the creation of a direct link. In formal terms:

$$S_{i,j}^{[n,n+1]} = q_i^{[n]} \cdot q_j^{[n+1]} \cdot e^{-\frac{d_{i,j}}{\alpha}} \cdot \frac{1 + \left(q_i^{[n+1]} - q_i^{[n]}\right)}{1 + \left(q_j^{[n+1]} - q_j^{[n]}\right)}; q \in [0,1];$$

$$S_{Win,j}^{[n,n+1]} = \underset{i}{ArgMax}\left\{S_{i,j}^{[n,n+1]}\right\};$$

$$\left\{\begin{array}{ll} C_{i,j}^{[n,n+1]} = 1 & i = Win; \\ C_{i,j}^{[n,n+1]} = 0 & i \neq Win; \end{array}\right\}.$$

Legend:

$q_i^{[n]}, q_i^{[n+1]}$ = Quantity in source place at the time n or at the time n + 1;

$q_j^{[n]}, q_j^{[n+1]}$ = Quantity in destination place at the time n or at the time n + 1;

$d_{i,j}$ = Distance between source and destination;

α = Tuned parameter connected to the distance;

$S_{i,j}^{[n,n+1]}$ = Strength of directed connection between source

at time n and destination at time n + 1;

$S_{Win,j}^{[n,n+1]}$ = Selection of the strongest connection between source

at time n and destination at time n + 1;

$C_{i,j}^{[n,n+1]}$ = Presence of a directed link between source

at time n and destination at time n + 1.

Table 8.2 An example

Data	Quantity at $n = 0$	Quantity at $n = 1$	Quantity at $n = 2$
Place 1	0	0	5
Place 2	0	0	2
Place 3	0	4	0
Place 4	1	3	7
Place 5	1	1	4

Fig. 8.1 The spatial distribution of entities as of Table 8.2

In other words, TDM reconstructs a causal scheme by singling out the 'main channel' through which diffusion takes place and by highlighting the corresponding links that carry out the basic channeling. To better elucidate the sense and the implication of this simple model, let us consider a specific, toy example. Consider the following table, providing us the data that describe the specific diffusion process (Table 8.2): with the corresponding spatial distribution of entities as in Fig. 8.1:

By applying the TDM equations, we easily derive the following computations for the transition from $n = 0$ to $n = 1$ (see Tables 8.3 and 8.4):

and see Tables 8.5 and 8.6 for the transition between $n = 1$ and $n = 2$.

In terms of causation schemes, we then have, in the transition from $n = 0$ to $n = 1$ (see Table 8.7)

and accordingly, in the transition from $n = 2$ to $n = 3$ (Table 8.8).

Table 8.3 TDM matrix of strength of the connections between time 0 and time 1

Strength of connections between places (Row = Time (n) − Columns = Time (n + 1))

n = 0; n = 1	Place 1	Place 2	Place 3	Place 4	Place 5
Place 1	0	0	0	0	0
Place 2	0	0	0	0	0
Place 3	0	0	0	0	0
Place 4	0	0	1	0	0.008928
Place 5	0	0	0.089095	0.016203	0

Table 8.4 TDM matrix of the plausible connections between time 0 and time 1

Presence of connections between places (Row = Time (n) − Columns = Time (n + 1))

n = 0; n = 1	Place 1	Place 2	Place 3	Place 4	Place 5
Place 1	0	0	0	0	0
Place 2	0	0	0	0	0
Place 3	0	0	0	0	0
Place 4	0	0	1	0	1
Place 5	0	0	0	1	0

Table 8.5 TDM matrix of strength of the connections between time 1 and time 2

Strength of connections between places (Row = Time (n) − Columns = Time (n + 1))

n = 1; n = 2	Place 1	Place 2	Place 3	Place 4	Place 5
Place 1	0	0	0	0	0
Place 2	0	0	0	0	0
Place 3	0.010040	0.002334	0	1	0.072003
Place 4	0.006473	0.005317	0	0	0.010714
Place 5	0.000994	0.000029	0	0.005165	0

Table 8.6 TDM matrix of the plausible connections between time 1 and time 2)

Presence of connections between places (Row = Time (n) − Columns = Time (n + 1))

n = 1; n = 2	Place 1	Place 2	Place 3	Place 4	Place 5
Place 1	0	0	0	0	0
Place 2	0	0	0	0	0
Place 3	1	0	0	1	1
Place 4	0	1	0	0	0
Place 5	0	0	0	0	0

Table 8.7 Transformation from n = 0 to n = 1, according to TDM

Transf_0 − > 1

Source	TRF	Destination	Strength
Place_4	→	Place_3	1
Place_5	→	Place_4	0.016203
Place_4	→	Place_5	0.008928

Table 8.8 Transformation from n = 1 to n = 2, according to TDM

Transf_1 − > 2

Source	TRF	Destination	Strength
Place_3	→	Place_1	0.010040
Place_4	→	Place_2	0.005317
Place_3	→	Place_4	1
Place_3	→	Place_5	0.072003

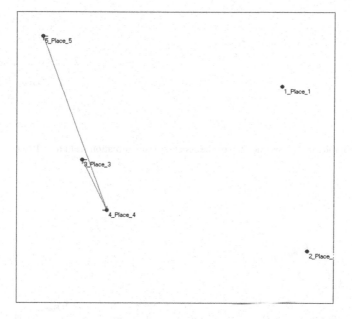

Fig. 8.2 Graphs for the first step of the process: transformation from n = 0 to n = 1

The above computations may be easily translated into graphs, thus obtaining the following for the two above transitions (respectively Figs. 8.2 and 8.3 and thus, by overlapping the first two steps, we obtain the following graph in Fig. 8.4).

We can of course proceed with further steps, thereby obtaining a dynamic representation of the direction of the main causation scheme of the process in terms of a graph of increasing complexity.

Now, in order to transform our discrete diffusion model into a proper diffusion process over a scalar field, we have to determine how the quantities located at each entity in each given moment influence the state of any given point of the scalar field with generic coordinates (x,y).

First of all, we determine the potential influence of each given entity at a given time by adding up the strength of its connection toward all other entities at that time,

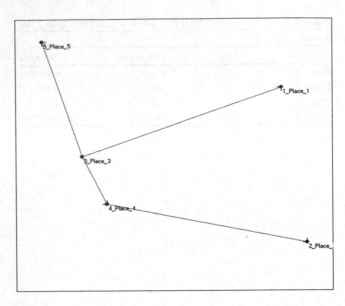

Fig. 8.3 Graphs for the second step of the process: transformation from n = 1 to n = 2

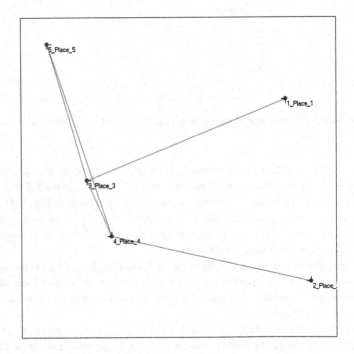

Fig. 8.4 Cumulative influence graph

and the cumulative potential influence of the entity by adding up potential influences over all of the time periods:

$$I_{i_{x,y}}^{[n,n+1]} = \sum_{j}^{M} S_{i,j}^{[n,n+1]}; \text{ Potential Influence of the i-th assigned entity}$$

upon the whole environment space;

$$CI_{i_{x,y}} = \sum_{n}^{N} \sum_{j}^{M} S_{i,j}^{[n]}; \text{ Cumulative Potential Influence of the i-th assigned entity}$$

upon the whole environment space;

$M = $ Number of the assigned entities;

$N = $ Number of time steps.

Next, we define the potential U of the scalar field through an (exponential) weighting of the potential influence of each given entity, controlled by the distance between any point of the field and that entity, and then adding over all of the entities:

$$U_{k_{x,y}}^{[n,n+1]} = \sum_{i}^{M} e^{-\frac{D\left(E_{i_{x,y}},P_{k_{x,y}}\right)}{\alpha}} \cdot I_{i_{x,y}}^{[n,n+1]};$$

$D(\cdot) = $ Distance between the generic k-th point (P) and the i-th assigned entity (E)

$M = $ Number of the assigned entities

Each given point of the field then *absorbs* some influence from each of the entity according to their relative strengths and distances.

We are thus in the position to define the corresponding scalar fields for the various steps of the diffusion process of our toy example thereby obtaining the following pictures (see Figs. 8.5 and 8.6).

And thus, the overlap of the two fields for each single influence step, following the same logic that yielded the cumulative influence graph as of Fig. 8.4, provides the actual scalar field of the influence after the first two steps (Fig. 8.7):

In its essence, we have thus described TDM: It is a diffusion model based on a gravitational logic which determines selectively the actual pattern of the influence lines that best describe a certain diffusion model among a certain finite number of entities, and can be extended in a natural and consequential way to a continuous diffusion model where the potential that generates the scalar field is obtained applying the same gravitational logic to all points of the environment space. In the following sections, after having presented the basic concepts of Twisting Theory (TWT), we will put this model at work on a specific and demanding example.

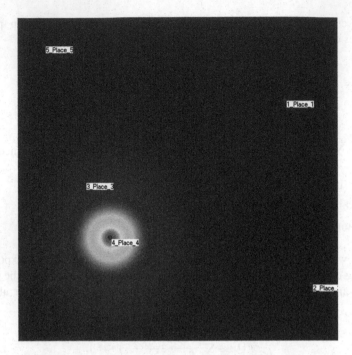

Fig. 8.5 Scalar field for the toy example: steps 1

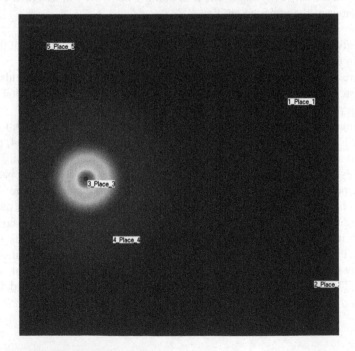

Fig. 8.6 Scalar field for the toy example: steps 2

Fig. 8.7 The scalar field with the overlap of the first two influence steps

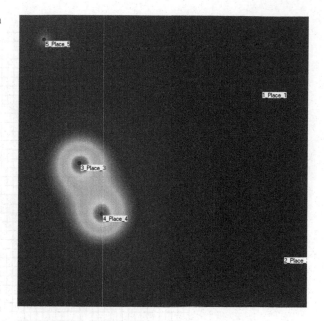

8.3   Twisting Theory and Twisting Algorithm (TWT)

With respect to the framework already defined in Sect. 8.2, imagine now that the entities located in the environmental space are no longer fixed but tend to move because of the action of some inherent force. In particular, this force can be interpreted as the quantity associated with the entity, which now gives a momentum to the entity that depends on the magnitude of the quantity itself. Each entity thus describes a trajectory in the environment space. For simplicity, we assume all such trajectories to be linear. Suppose moreover that the space is covered by a grid, and that the force that acts on the entities also acts on the grid, so that the change of position of the entities brings about accordingly a distortion of the original grid, i.e., the field of forces generated by the quantities assigned to the entities may now be visualized in terms of the modifications that it produces on the grid. We call all points belonging to the grid *geometrical points*. How can we describe the actual effect onto the grid of the forces acting upon the entities?

To illustrate the idea, consider the following 36 × 36 grid with five entities and a trajectory in two time steps (Fig. 8.8).

The approach of TWT is to divide trajectories into N given sub-steps of equal length, so that each entity is dynamically identified by the coordinates of its place of

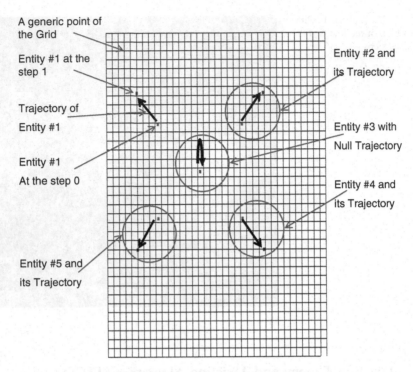

Fig. 8.8 An example of entity trajectories across the grid

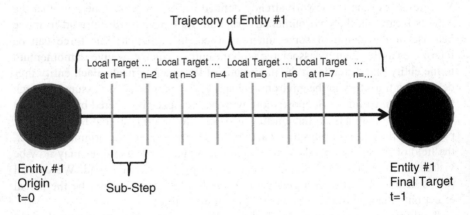

Fig. 8.9 Splitting of the entity trajectory into sub-steps

origin, by the moving local target corresponding to the movement at each single sub-step, and by the coordinates of the place of arrival of the trajectory (Fig. 8.9).

TWT focuses upon the distances between a given geometrical point (i.e., a point of the grid) and the position of each given entity as it moves along its trajectory, measuring how such distance varies at each sub-step of the trajectory.

Fig. 8.10 The sub-trajectory of the entity and the accumulation of potential energy

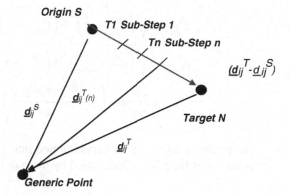

$$d_{ij}^S(n) = \sqrt{\left(x_i^P(n) - x_j^S\right)^2 + \left(y_i^P(n) - y_j^S\right)^2}$$

$$d_{ij}^T(n) = \sqrt{\left(x_i^P(n) - x_j^T(n)\right)^2 + \left(y_i^P(n) - y_j^T(n)\right)^2}.$$

Legend :

$x_i^P(n), y_i^P(n) = $ Coordinates of a generic point (i) of the grid at substep (n); when n = 0, geometrical points all sit upon the regular grid.

$x_j^S, y_j^S = $ Origin Coordinates of each entity (j).

$x_j^T(n), y_j^T(n) = $ Local target of the coordinates of each entity (j) at any substep (n); when n = 0, the entity lies at its Origin, whereas when n = N the entity has completed its trajectory.

$d_{ij}^S(n) = $ Distance of a generic point (i) from the Origin of any entity at substep (n);

$d_{ij}^T(n) = $ Distance of a generic point (i) from the Local Target (n) of any entity

atsubstep(n);

This variation in the distance can be interpreted as the accumulation of potential energy that is consequently free to act upon the grid, i.e. upon the position of the geometrical points themselves. The more the varying position of the entity along its trajectory tends to change its distance from the given geometrical point from step to step, the more the grid is consequently 'stretched' or 'shrunk' accordingly (Fig. 8.10).

In mathematical terms, the variation of potential energy Δ across the whole trajectory may be thus expressed as:

$$d^S_{i,j_{(n)}} = \sqrt{\left(x^P_{i_{(n)}} - x^S_j\right)^2 + \left(y^P_{i_{(n)}} - y^S_j\right)^2}$$

$$d^T_{i,j_{(n)}} = \sqrt{\left(x^P_{i_{(n)}} - x^T_{j_{(n)}}\right)^2 + \left(y^P_{i_{(n)}} - y^T_{j_{(n)}}\right)^2}$$

$$\Delta_{i_{(n)}} = \sum_{j=1}^{N} \exp\left(-\frac{\left(d^S_{i,j_{(n)}} + d^T_{i,j_{(n)}}\right)}{\alpha}\right) \cdot \left|d^S_{i,j_{(n)}} - d^T_{i,j_{(n)}}\right|.$$

The potential energy is then transformed into kinetic energy, i.e. it produces a deformation of the grid as determined by the movement of the entity:

$$\delta x^{[p]}_{i_{(n)}} = \sum_{j}^{N} \exp\left(-\frac{d^{[s]}_{i,j_{(n)}} + d^{[t]}_{i,j_{(n)}}}{\alpha}\right) \cdot \left(x^{[s]}_j - x^{[t]}_{j_{(n)}}\right);$$

$$\delta y^{[p]}_{i_{(n)}} = \sum_{j}^{N} \exp\left(-\frac{d^{[s]}_{i,j_{(n)}} + d^{[t]}_{i,j_{(n)}}}{\alpha}\right) \cdot \left(y^{[s]}_j - y^{[t]}_{j_{(n)}}\right);$$

$$x^{[p]}_{i_{(n+1)}} = x^{[p]}_{i_{(n)}} + \Delta_{i_{(n)}} \delta x^{[p]}_{i_{(n)}} < 0;$$

$$x^{[p]}_{i_{(n+1)}} = x^{[p]}_{i_{(n)}} - \Delta_{i_{(n)}} \delta x^{[p]}_{i_{(n)}} \geq 0;$$

$$y^{[p]}_{i_{(n+1)}} = y^{[p]}_{i_{(n)}} + \Delta_{i_{(n)}} \delta y^{[p]}_{i_{(n)}} < 0;$$

$$y^{[p]}_{i_{(n+1)}} = y^{[p]}_{i_{(n)}} - \Delta_{i_{(n)}} \delta y^{[p]}_{i_{(n)}} \geq 0.$$

We are now in the position to check how, by superimposing the TWT framework upon the TDM-focused example introduced earlier, it is possible to visualize the diffusion process in terms of the deformation of the corresponding grid, by simply applying the equations introduced above. Switching back to the step 1 of the process, and recalling the graph that describes the influence pattern, we can accordingly derive the associated Twisting Map (see Figs. 8.11 and 8.12):

Accordingly, we can do the same for the influence graph at step 2 (Figs. 8.13 and 8.14), and for the cumulative influence graph, thus obtaining the actual Twisting Map after two steps of the process (Figs. 8.15 and 8.16).

If our conceptualization of the diffusion process then turns out to have some sense, by the help of TDM we should have been able to determine the actual causal processes that act upon entities and bring about the evolution of the quantity values from one step to another, and by TWT we should have been able to represent how the whole causal dynamics would bring about a complex, nonlinear field dynamics of diffusion even starting from a simple interpretation of quantities associated to each entity as linear impulses. To check whether our intuition is reasonably grounded, we will have to evaluate it against a benchmark that allows us to test it severely enough.

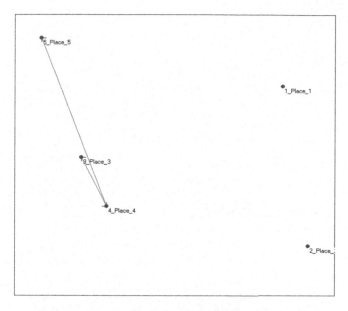

Fig. 8.11 Toy example: the influence graph at step 1 according to TDM

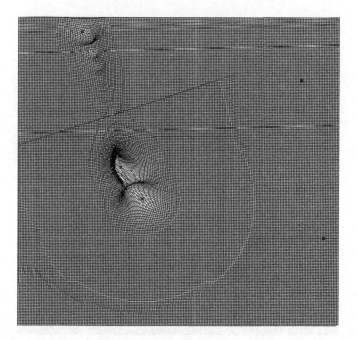

Fig. 8.12 Toy example: the influence graph at step 1 coded by Twisting Theory

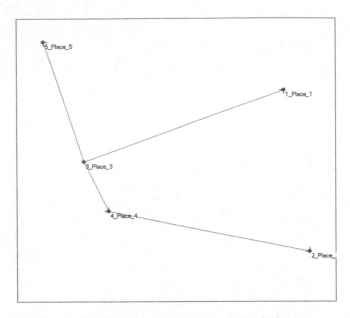

Fig. 8.13 Toy example: the influence graph at step 2 according to TDM

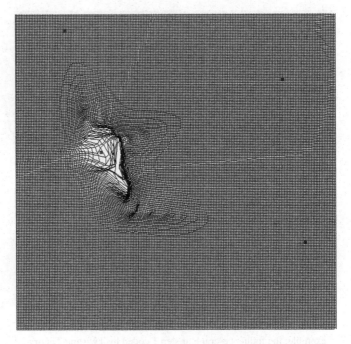

Fig. 8.14 Toy example: the influence graph at step 2 coded by Twisting Theory

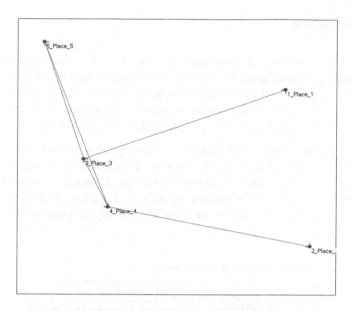

Fig. 8.15 Toy example: the influence graph after steps 1–2 according to TDM

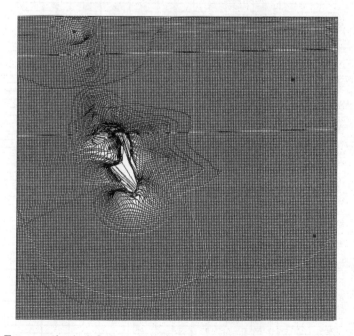

Fig. 8.16 Toy example: the influence graph after steps 1–2 and the corresponding Twisting Map

8.4 Prediction

In order to proceed with the construction of our test, we now begin by slightly complicating our toy example by introducing two extra steps in our time series:

Assume now that, on the basis of this expanded database, we want to make some one-step prediction on the values of the quantities assigned to each entity. There are at least two reasons that make this task pretty difficult (Table 8.9):

- The number of records in the dataset is extremely small (five observations), and even more so with reference to the number of variables (five variables);
- The prediction of an integer number (which has to do with the computation of an approximation function) is much more difficult than, say, the recognition of a pattern (i.e., an exercise in classification that is a standard test bed for machine learning).

Table 8.9 Expanded time series for the toy example

ID	Place Name	Lat and Long Y	X	n=0	n=1	n=2	n=3	n=4	n=5	n=6	Total
1	POPULINA	-50.5300	-19.9500	0	0	0	0	19	12	1	32
2	OUROESTE	-50.3700	-20.0000	0	0	2	6	5	0	0	13
3	INDIAPORA	-50.2900	-19.9800	0	0	0	0	5	0	0	5
4	PAULO DE FARIA	-49.3800	-20.0300	1	30	67	43	40	14	1	196
5	CARDOSO	-49.9100	-20.0800	1	1	4	2	4	1	0	13
6	RIOLANDIA	-49.6800	-19.9900	1	2	11	38	24	9	2	87
7	TURMALINA	-50.4700	-20.0500	0	0	0	0	1	0	0	1
8	MACEDONIA	-50.1900	-20.1400	0	0	0	2	3	1	0	6
9	DOLCINOPOLIS	-50.5100	-20.1200	0	0	0	0	0	1	0	1
10	PEDRANOPOLIS	-50.1100	-20.2400	0	0	2	0	0	0	0	2
11	FERNANDOPOLIS	-50.2400	-20.2800	0	11	73	205	201	35	17	542
12	ESTRELA D'OESTE	-50.4000	-20.2800	0	1	0	3	2	2	0	8
13	S. FE DO SUL	-50.9200	-20.2100	0	0	0	5	5	1	1	12
14	JALES	-50.5400	-20.2600	0	1	5	16	4	1	0	27
15	GUARACI	-48.9400	-20.4900	0	0	1	6	4	0	0	11
16	AMERICO DE CAMPOS	-49.7300	-20.2900	0	0	0	0	1	0	0	1
17	ICEM	-49.1900	-20.3400	0	0	0	1	3	0	2	6
18	COSMORAMA	-49.7700	-20.4700	0	0	0	2	4	3	1	10
19	NOVA GRANADA	-49.3100	-20.5300	0	0	1	3	4	0	0	8
20	MERIDIANO	-50.1700	-20.3500	0	0	0	1	0	0	0	1
21	VOTUPORANGA	-49.9700	-20.4200	2	4	9	7	10	3	4	39
22	TANABI	-49.6400	-20.6200	0	0	0	0	12	5	1	18
23	APARECIDA D'OESTE	-50.8800	-20.4400	0	0	1	0	0	0	0	1
24	GENERAL SALGADO	-50.3600	-20.6400	0	0	0	7	17	3	1	28
25	AURIFLAMA	-50.5500	-20.6800	0	0	0	0	3	0	0	3
26	IPIGUA	-49.3800	-20.6500	0	0	0	0	4	0	0	4
27	OLIMPIA	-48.9100	-20.7300	1	8	55	242	356	98	12	772
28	MONTE APRAZIVEL	-49.7100	-20.7700	0	0	6	20	18	18	0	62
29	GUAPIACU	-49.2200	-20.7900	0	0	1	2	5	2	2	12
30	BALSAMO	-49.5800	-20.7300	0	0	0	0	0	2	0	2
31	SAO JOSE DO RIO PRETO	-49.5200	-20.8200	30	142	583	2002	2781	690	188	6416
32	POLONI	-49.8200	-20.7800	0	0	0	2	12	9	3	26
33	MIRASSOL	-49.5200	-20.8100	2	21	171	550	618	159	44	1565
34	MACAUBAL	-49.9600	-20.8000	0	0	0	1	2	2	0	5
35	SEVERINIA	-48.8000	-20.8000	0	0	0	1	4	0	0	5
36	NEVES PAULISTA	-49.6300	-20.8400	0	0	0	13	32	11	2	58
37	UCHOA	-49.1700	-20.9500	0	0	0	0	2	6	2	10
38	TABAPUA	-49.0300	-20.9600	10	23	7	6	3	2	0	51
39	CAJOBI	-48.8000	-20.8800	2	2	84	179	39	1	0	307
40	CEDRAL	-49.2600	-20.9000	0	0	0	0	3	2	1	6
41	JACI	-49.5700	-20.8800	0	0	1	5	6	2	0	14
42	BADY BASSITT	-49.4400	-20.9100	0	0	1	8	11	6	2	28
43	EMBAUBA	-48.8300	-20.9800	0	0	0	1	1	0	0	2
44	JOSE BONIFACIO	-49.6800	-21.0500	0	0	1	20	15	9	6	51
45	IBIRA	-49.2400	-21.0800	0	0	0	0	1	0	0	1
46	POTIRENDABA	-49.3700	-21.0400	0	0	0	0	1	5	1	7
47	CATANDUVA	-48.9700	-21.1300	0	0	14	38	43	14	2	111
48	CATIGUA	-49.0500	-21.0400	0	1	20	41	6	1	0	69
49	PALM. PAULISTA	-48.8000	-21.0800	0	0	0	1	1	0	0	2
50	URUPES	-49.2900	-21.2000	0	0	0	0	7	2	0	9
51	IRAPUA	-49.4000	-21.2700	0	0	2	0	1	0	0	3
52	ADOLFO	-49.6400	-21.2300	0	0	0	1	1	0	0	2
53	SANTA ADELIA	-48.8000	-21.2400	0	0	0	0	1	3	0	4
54	ITAJOBI	-49.0500	-21.3100	0	0	0	1	0	0	0	1

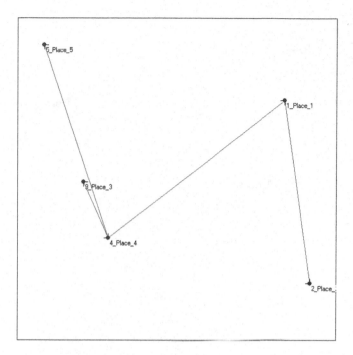

Fig. 8.17 Graphs for the third step of the process

As a consequence, under these conditions it is virtually impossible, from a statistical viewpoint, to construct a viable prediction model, and even more so if this model calls upon some Artificial Neural Network (ANN) architecture. At this point, however, we are in the position to appreciate the potential of TDM to carry out such an apparently impossible task.

By means of the TDM procedure, we can transform the small original dataset into a set of four 5×5 matrices, containing the strengths of the connections between steps n and $n + 1$ of the influence process; let us call them the matrices of Strength S. Moreover, we can accordingly construct another set of companion matrices, delivering the oriented links among entities from step n to step $n + 1$; let us call them the matrices of Connections C (see Data Appendix: Tables A.1, A.2 and A.3). In addition to the two graphs for steps 1 and 2 already shown in Figs. 8.2 and 8.3 above, we thus obtain the following graphs for steps 3 and 4 respectively (see Figs. 8.17 and 8.18).

At this point, the Connection matrices C generated by the TDM procedure can be stacked into a common dataset with a mobile window, so that connections that refer to a same entity at different times are lined up. We can thus generate a substantially bigger dataset, which can be learned by a suitably designed supervised ANN:

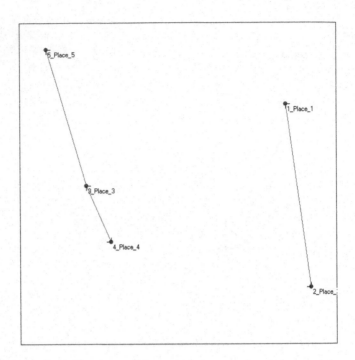

Fig. 8.18 Graphs for the fourth step of the process

$$\text{New Dataset}: \left\{ \left\{ \mathbf{x}_n^p, \mathbf{x}_{n+1}^p \right\}_{n=0}^{N-1} \right\}_{p=1}^{M}$$

Legend :

M = Number of entities;

N = Number of time steps;

consequently:

ANN Model: $x(n) \rightarrow x(n+1) = f(x(n), w^*) + \varepsilon.$

The ANN weights matrix w^* that is obtained once the training phase has been completed should thus be able to encode the whole dynamic process described by the time series, by reconstructing it from the local transition laws that have been the object of learning. In Table 8.10, we show synthetically how the available data are divided into the Training, Testing and Prediction datasets, respectively.

Thus, for the Training and Testing sets, each record is made up of $M + 1$ input variables: The connectivity of each entity from time n to time $n + 1$, as determined by the TDM algorithm, and an identification number (i.e., an integer) for each entity (position). The M target variables, consequently, will be the connectivity of each entity from time $n + 1$ to time $n + 2$.

Overall, the Training set if made up of 10 patterns, each one with 6 input and 5 target variables. The Testing set contains 5 patterns with analogous characteristics:

Table 8.10 Training, testing and prediction datasets for the stacked connection matrices of the toy problem

TDM Algorithm	
Working Hypothesis	378
Right Connections	183
Missing Connections	3
Sensitivity	98.39%
Specificity	100.00%
Global Accuracy	99.21%

the dataset still remains rather small. The ANN that will be trained is a Sine Net[3] (Buscema 2000; Buscema et al. 2006). Once training is over, it will be tested in a blind way, i.e., it will only be shown the input variables in order to predict the outputs. The ANN will thus make a total of 25 independent predictions (5 output variables for 5 rounds).

The predictive performance of the Sine Net ANN will be measured according to three different standards: sensitivity, namely, the number of actual connections correctly predicted; specificity, namely, the number of actual non-connections correctly predicted; and accuracy, namely, the overall number of correct predictions.

The chosen specification of the Sine Net has 24 hidden units, and its training has been terminated after 43989 epochs, with a RMSE of 0.21505214. The results of the testing are shown in Table 8.11.

As it can be noticed, the results are pretty interesting; even in a prohibitive learning condition as the present one, the Sine Net reaches an accuracy of 0.84, and a specificity of 0.9. Sensitivity stops at 0.6, but even in this case it remains well above the critical threshold of 0.5. Figure 8.19 reports the comparison between the actual and the predicted dynamics at the last time step.

What made this test particularly severe, however, was the fact that, in the toy problem, data were basically picked up at random, i.e. they did not reflect any implicit, coherent semantics. In spite of this, the ANN was able to find out a substantial part of the structural properties of the process by means of a simple application of the TDM procedure.

We now test the methodology against a more rich and challenging example, and in particular one where there actually is a strong implicit semantics. This last feature invites us to think that, if our methodology is sound enough, we should expect a *better* performance than for the toy problem, in spite of the much greater complexity of the underlying diffusion process, in that we can now capitalize upon the far richer semantics of the problem, which should now possibly reveal many structural layers about which the ANN can learn proficiently.

---

[3] Sine Net is a USA Patent No. US 7,788,196 B2 (Aug. 31, 2010), inventor: Massimo Buscema, m. buscema@semeion.it; owners Semeion Research Center of Sciences of Communication, via Sersale 117, Rome, 00128, Italy.

Table 8.11 Results of the testing: Missing connections have a dotted background, false connections have horizontal ruling

Blind Test	Place(n+1) 1	Place(n+1) 2	Place(n+1) 3	Place(n+1) 4	Place(n+1) 5	
Place(n) 1	0	1	0	0	0	Real
	0	1	1	0	0	Estimation
Place(n) 2	1	0	0	0	0	Real
	0	0	0	0	0	Estimation
Place(n) 3	0	0	0	1	1	Real
	1	0	0	1	1	Estimation
Place(n) 4	0	0	1	0	0	Real
	0	3	0	0	0	Estimation
Place(n) 5	0	0	0	0	0	Real
	0	0	0	0	0	Estimation
Sensitivity: 3/5 ->	60.00%	Specificity: 17/20 ->		85.00%		Accuracy: 20/25 80.00%

Sn						
Blind Test	Place(n+1) 1	Place(n+1) 2	Place(n+1) 3	Place(n+1) 4	Place(n+1) 5	
Place(n) 1	0	1	0	0	0	Real
	0	1	1	0	0	Estimation
Place(n) 2	1	0	0	0	0	Real
	0	0	0	0	0	Estimation
Place(n) 3	0	0	0	1	1	Real
	1	0	0	1	1	Estimation
Place(n) 4	0	0	1	0	0	Real
	0	0	0	0	0	Estimation
Place(n) 5	0	0	0	0	0	Real
	0	0	0	0	0	Estimation
Sensitivity: 3/5 ->	60.00%	Specificity: 18/20 ->		90.00%		Accuracy: 21/25 84.00%

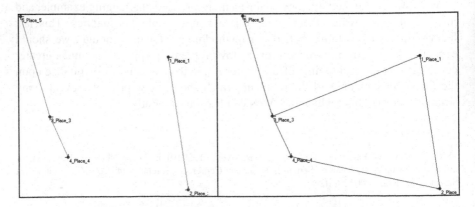

Fig. 8.19 Real (*left*) versus predicted (*right*) dynamics for the Sine Net

8.5 A Real World Test: The 2001 Dengue Fever Epidemics in Brazil

8.5.1 The Dengue Fever Epidemics Case Study

This section applies our methods on an actual case, the case of Dengue fever epidemic in the state of São Paulo, Southeast Brazil, as reported in Ferreira (2010). The map of the epidemics diffusion is reported in Fig. 8.20, and the data about the actual pattern of diffusion of contagion over 7 months in the 54 counties where at least one case was reported are listed in Table 8.12. As one can see from the table, the spatial coordinates of the main center of each county have been determined, and time steps span one month each. At each step, the associated quantity corresponds to the number of new infected subjects. From these data, a digital map has been constructed, which reports the main centers of all of the 54 counties that have witnessed at least one case of the fever epidemics, reported in Fig. 8.21. This map then constitutes our set of entities, which we use to analyze the actual propagation dynamics, according to the methods presented in the previous sections.

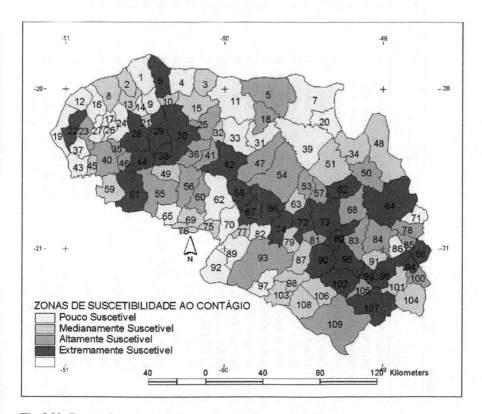

Fig. 8.20 Dengue fever epidemics: map of contagion susceptibility

Table 8.12 Contagion data over a 7-months period for 54 regions

ID	Place name	Lat and Long		Number of new subjects infected at each step							Total
		Y	X	n = 0	n = 1	n = 2	n = 3	n = 4	n = 5	n = 6	
1	POPULINA	−50.5300	−19.9500	0	0	0	0	19	12	1	32
2	OUROESTE	−50.3700	−20.0000	0	0	2	6	5	0	0	13
3	INDIAPORA	−50.2900	−19.9800	0	0	0	0	5	0	0	5
4	PAULO_DE_FARIA	−49.3800	−20.0300	1	30	67	43	40	14	1	196
5	CARDOSO	−49.9100	−20.0800	1	1	4	2	4	1	0	13
6	RIOLANDIA	−49.6800	−19.9900	1	2	11	38	24	9	2	87
7	TURMALINA	−50.4700	−20.0500	0	0	0	0	1	0	0	1
8	MACEDONIA	−50.1900	−20.1400	0	0	0	2	3	1	0	6
9	DOLCINOPOLIS	−50.5100	−20.1200	0	0	0	0	0	1	0	1
10	PEDRANOPOLIS	−50.1100	−20.2400	0	0	2	0	0	0	0	2
11	FERNANDOPOLIS	−50.2400	−20.2800	0	11	73	205	201	35	17	542
12	ESTRELA_D'OESTE	−50.4000	−20.2800	0	1	0	3	2	2	0	8
13	S_FE_DO_SUL	−50.9200	−20.2100	0	0	0	5	5	1	1	12
14	JALES	−50.5400	−20.2600	0	1	5	16	4	1	0	27
15	GUARACI	−48.9400	−20.4900	0	0	1	6	4	0	0	11
16	AMERICO_DE_CAMPOS	−49.7300	−20.2900	0	0	0	0	1	0	0	1
17	ICEM	−49.1900	−20.3400	0	0	0	1	3	0	2	6
18	COSMORAMA	−49.7700	−20.4700	0	0	0	2	4	3	1	10
19	NOVA_GRANADA	−49.3100	−20.5300	0	0	1	3	4	0	0	8
20	MERIDIANO	−50.1700	−20.3500	0	0	0	1	0	0	0	1
21	VOTUPORANGA	−49.9700	−20.4200	2	4	9	7	10	3	4	39
22	TANABI	−49.6400	−20.6200	0	0	0	0	12	5	1	18
23	APARECIDA_D'OESTE	−50.8800	−20.4400	0	0	1	0	0	0	0	1
24	GENERAL_SALGADO	−50.3600	−20.6400	0	0	0	7	17	3	1	28
25	AURIFLAMA	−50.5500	−20.6800	0	0	0	0	3	0	0	3

No.	Name	Long.	Lat.								
26	IPIGUA	−49.3800	−20.6500	0	0	0	0	4	0	0	4
27	OLIMPIA	−48.9100	−20.7300	1	8	55	242	356	98	12	772
28	MONTE_APRAZIVEL	−49.7100	−20.7700	0	0	6	20	18	18	0	62
29	GUAPIACU	−49.2200	−20.7900	0	0	1	2	5	2	2	12
30	BALSAMO	−49.5800	−20.7300	0	0	0	0	0	2	0	2
31	SAO_JOSE_DO_RIO_PRETO	−49.3700	−20.8200	30	142	583	2002	2781	690	188	6416
32	POLONI	−49.8200	−20.7800	0	0	0	2	12	9	3	26
33	MIRASSOL	−49.5200	−20.8100	2	21	171	550	618	159	44	1565
34	MACAUBAL	−49.9600	−20.8000	0	0	0	1	2	2	0	5
35	SEVERINIA	−48.8000	−20.8000	0	0	0	1	4	0	0	5
36	NEVES_PAULISTA	−49.6300	−20.8400	0	0	0	13	32	11	2	58
37	UCHOA	−49.1700	−20.9500	0	0	0	0	2	6	2	10
38	TABAPUA	−49.0300	−20.9600	10	23	84	6	3	2	0	51
39	CAJOBI	−48.8000	−20.8800	2	2	0	179	39	1	0	307
40	CEDRAL	−49.2600	−20.9000	0	0	1	0	3	2	1	6
41	JACI	−49.5700	−20.8800	0	0	1	5	6	2	0	14
42	BADY_BASSITT	−49.4400	−20.9100	0	0	1	8	11	6	2	28
43	EMBAUBA	−48.8300	−20.9800	0	0	0	1	1	0	0	2
44	JOSE_BONIFACIO	−49.6800	−21.0500	0	0	1	20	15	9	6	51
45	IBIRA	−49.2400	−21.0800	0	0	0	0	1	0	0	1
46	POTIRENDABA	−49.3700	−21.0400	0	0	0	0	1	5	1	7
47	CATANDUVA	−48.9700	−21.1300	0	0	14	38	43	14	2	111
48	CATIGUA	−49.0500	−21.0400	0	1	20	41	6	1	0	69
49	PALM._PAULISTA	−48.8000	−21.0800	0	0	0	1	1	0	0	2
50	URUPES	−49.2900	−21.2000	0	0	2	0	7	2	0	9
51	IRAPUA	−49.4000	−21.2700	0	0	0	0	1	0	0	3
52	ADOLFO	−49.6400	−21.2300	0	0	0	1	1	0	0	2
53	SANTA_ADELIA	−48.8000	−21.2400	0	0	0	1	3	0	0	4
54	ITAJOBI	−49.0500	−21.3100	0	0	0	1	0	0	0	1

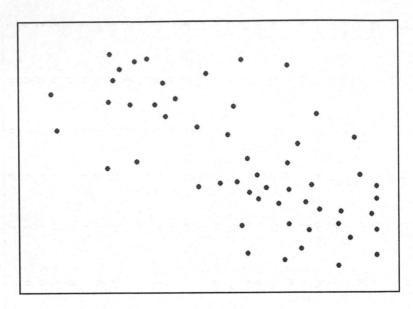

Fig. 8.21 Digital map of the outbreak

8.6 The Harmonic Center[4]: A New Concept and Mathematical Entity

Even without exploiting the techniques for the analysis of the dynamics of diffusion of the epidemics presented in this paper, a good deal of information can be extracted already by the analysis of the static distribution of points. Notice, moreover, that the static distribution says nothing about the actual level of susceptibility of the various counties, i.e., does not differentiate among the points that carry a high number of infected subjects and those that carry very little ones and in the limit, just one. Only on the basis of this digital map, however, through the further development of techniques presented and applied in Buscema et al. (2009), we are able to locate with a certain precision the point of maximal condensation of the epidemics by calculating the so called harmonic center of the scatter of points. It is interesting to stress that the harmonic center does not coincide in principle with the outbreak point of the epidemics (whose estimate is better found with a complementary method, the TWC [see below]) – it is rather the point where the dynamic impulse that drives the propagation of the epidemics exerts its most significant impact. For intuitive reasons, in the case of epidemics the outbreak and the harmonic center tend to be

---

[4] Harmonic Center is a USA Patent pending # 12/969,620 (16-Dec-2010), inventor: Massimo Buscema, m.buscema@semeion.it; owners Semeion Research Center of Sciences of Communication, via Sersale 117, Rome, 00128, Italy and CSI Research & Environment, via CesarePavese, 305, Rome, Italy.

close by, in that it is around the outbreak that the epidemics generally unleashes its strongest virulence. For other diffusion phenomena such as earthquakes or massive landslides, however, the epicenter and the harmonic center may be even quite far away from each other.

One also can see the harmonic center as the point from where one accumulates the influence of each given entity, as defined in Sect. 8.2, is maximized.

In particular, the harmonic point is the perfect phasing point such that the squared distances from the various entities are as close as possible to being exact multiples of one another (i.e., they are harmonics). In principle, one can partition the environmental space into regions according to their degree of harmony. To make a simple association, we could identify harmony with temperature, and then, those points which are more harmonic with respect to each assigned entity can be said to be 'hot' (max harmony). The 'warm' points (high harmony) are those whose harmony is below that of some entities but still more harmonic that the average of the entities. The 'cool' points (low harmony) are those lying among the average harmony of the points of the environment space and that of the entities, whereas the 'cold' points (null harmony) are those with harmony below the average of the environment space.

We formally denote entities as 'assigned points' and 'generic points' of the environmental space as 'pixel points' and calculate distances between all of the assigned points and each specific pixel point in the process of calculating their harmony with reference to the harmony function dH. Once the value of dH has been obtained, the environmental space is portioned. The method is as follows:

Harmonic Centre Calculation:

$$d_{k,i} = \left\lfloor C \cdot \left(1 + (x_k - x_i)^2 + (y_k - y_i)^2\right)\right\rfloor;$$

$$dH_k = 1.0 - f\left(\frac{\sum\limits_{i=1}^{N}\sum\limits_{j=1,j\neq i}^{N} Rem\left(d_{k,i}, d_{k,j}\right)}{\frac{1}{M}\cdot\sum\limits_{k=1}^{M}\sum\limits_{i=1}^{N}\sum\limits_{j=1,j\neq i}^{N} Rem\left(d_{k,i}, d_{k,j}\right)}\right);$$

$$dHarmonic = \arg\max_{k,k\in A\cup P}\{dH_k\};$$

Harmonic Field Segmentation:

$$d\bar{H}_M = \frac{1}{M}\cdot\sum\limits_{k=1,k\in P}^{M} dH_k;$$

$$d\bar{H}_N = \frac{1}{N}\cdot\sum\limits_{k=1,k\in A}^{N} dH_k;$$

$$dH_{MAX_N} = \max_{k,x_k\in A}\{dH_k\};$$

$$dH_k \in H_{Max} dH_k > H_{MAX};$$

$$dH_k \in H_{High} d\bar{H}_N < dH_k \leq H_{MAX};$$

$$dH_k \in H_{Low} d\bar{H}_M < dH_k \leq d\bar{H}_N;$$

$$dH_k \in H_{Null} dH_k < d\bar{H}_M.$$

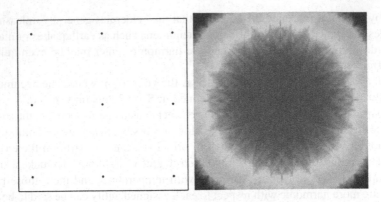

Fig. 8.22 The harmonic point and the harmonic map

where:

$N =$ Number of Assigned Points;

$M =$ Number of Pixel Points;

$A =$ set of Assigned Points;

$P =$ set of Pixel Points;

$i, j \in \{1, 2, ..., N\}$;

$k \in \{1, 2, ..., M\}$;

$C =$ big integer positive Constant;

$d =$ Euclidean Distance

$dH =$ Harmonic Distance;

$f() =$ Linear scaling between [0,1];

$dHarmonic =$ Harmonic Centre;

$d\bar{H}_N =$ Harm. Distance Mean of the Assigned Points;

$d\bar{H}_M =$ Harm. Distance Mean of the Pixel Points;

$dH_{MAX_N} =$ Minimal Harmonic Distance among the Assigned Points;

$H_{Max} =$ set of Points with Max Harmony;

$H_{High} =$ set of Points with High Harmony;

$H_{Low} =$ set of Points with Low Harmony;

$H_{Null} =$ set of Points with Null Harmony;

or the simple example of a 3 × 3 grid within a square, for instance, the harmonic center coincides with the central point of the grid (Fig. 8.22), and the geography of the harmonic regions turns out to be the following (with colors corresponding to 'temperatures'):

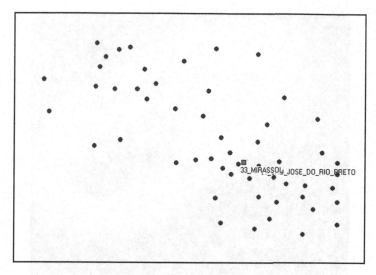

Fig. 8.23 The harmonic center of the distribution of points

In the case of the Dengue epidemics, the harmonic center is found to be between Mirassol and São Josè do Rio Preto, that is to say, between the actual outbreak point (São Josè) and its closest point of propagation (Mirassol). As expected, the estimated point of maximal virulence is close to the outbreak point, and in fact falls into an area with very high susceptibility according to the map elaborated by Ferreira (2010). The result is particularly striking, considering that the actual distribution of points does not give, as already remarked, any information about the actual number of infected subjects located in each point, but only refers to the semantics of space, i.e. to the structure of the reciprocal distances among the various points (Fig. 8.23).

8.7   The Topological Weight Centroid (TWC)

Remaining within a static context, i.e. without exploiting yet the potential of the TDM + TWT approach, we can however reconstruct a vector field for the propagation of the Dengue fever on the basis of the information contained in the digital map by constructing the so called TWC map, following the approach illustrated in Buscema and Grossi (2009), Chaps. 8 and 9. In this way, we give an estimate of the probability of diffusion of the epidemics from its outbreak, or better, from another close estimate of its actual position, called the Topologically Weighted Centroid (TWC). As it can be clearly read from Fig. 8.24, the harmonic center sits well within the area of maximal susceptibility estimated according to the TWC method, guaranteeing that the TWC and the harmonic center provide mutually coherent indications for this case, once again placing the outbreak and the peak point close to each other.

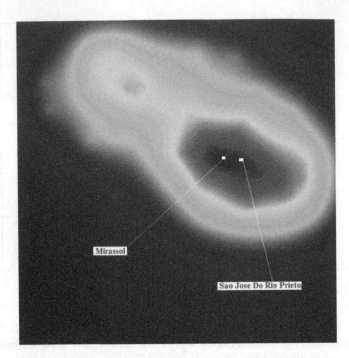

Mirassol

Sao Jose Do Rio Prieto

Fig. 8.24 TWC map of the Dengue fever epidemics

The TWC field, moreover, provides an interesting static benchmark reference for the further analysis that will be provided by means of the TDM + TWT approach.

Another important implication within the static informational framework comes from the analysis of the so-called TWC paths, which represent a 'nonlinear minimum spanning tree' for the spatial distribution of entities, i.e., the paths of minimum energy for the propagation dynamics. Quite interestingly, these paths replicate quite accurately the existing structure of communication networks (roads), as it can be checked in Fig. 8.25 which also can be seen from Google Earth aerial view of the region. As found out by Ferreira (2010), it turns out that the road network is actually the channel through which the epidemics has found its most efficient channeling, and then one can read the TWC paths map as a diagram of the propagation dynamics of the epidemics. But what is surprising once more is that in the digital map which has been the (static, non-quantified) informational basis for this computation has no reference whatsoever to the road network, and thus there is no way that the TWC algorithm could exploit this information in finding out its paths of minimal energy. Thus, in this perspective the structure of the road network emerges endogenously as the most efficient solution of the propagation problem for the epidemics.

Thus, quite interestingly, we conclude that, already from a static perspective and making use of techniques already introduced in previous work, it is already possible to make powerful inferences about the outbreaks and diffusion dynamics of the epidemics. To what extent the introduction of the dynamic TDM + TWT approach allows us to take a further step forward in the understanding of the diffusion phenomenon for this specific case?

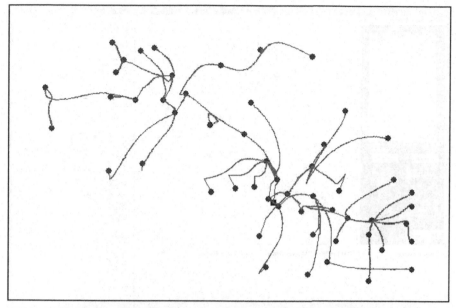

Fig. 8.25 *Above*: the TWC paths for the propagation of the Dengue fever epidemics. *Below*: aerial view from Google Earth

8.8 TDM Applied to the Dengue Fever Epidemics Data

To answer this question, we can simply apply the TDM methodology to the digital map introduced in Fig. 8.21 above for the Dengue fever epidemic, this time assigning to each entity the quantity corresponding to the number of newly infected subjects from time to time. The result of step 1 of the procedure is shown in

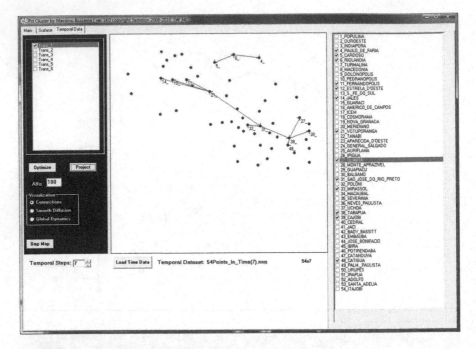

Fig. 8.26 The causation process of the epidemics, step 1

Fig. 8.26: the axis São Josè-Mirassol, linking the outbreak and the peak point, is identified, and we clearly single out two branches, a Northwest-bound one and a Southeast-bound one, through which the epidemics reach out to new counties. Interestingly, there is also a minor Northeastern branch that seems to have developed independently of the main one.

At this point, looking at the whole sequence of steps 1–6 (Fig. 8.27), it becomes clear how, starting from the first outreach run, the epidemic then develops according to a peculiar logic: The Northwestern branch, initially infected by carriers coming from the outbreak county, now becomes a secondary infectious hub of its own, gradually spreading over in nearby centers of the Northwestern quadrant and becoming causally independent from the outbreak from step 2 onwards. At the outbreak county, on the other hand, we witness the emergence of a 'double bind' connection between the outbreak and the peak points which signals a strong positive feedback causal loop, which causes the building of further virulence, with a very far-reaching and persistent propagation that remains at high levels over the months, gradually extending to once marginal places. The initially independent Northeastern branch of virulence is rapidly 'colonized' by the outbreak/peak infection hub and subsequently persists in this new status in spite of the relative distance from the hub itself and its being relatively closer to the secondary Northwestern hub – a further signal of the fact that the peak point actually stays close to the outbreak.

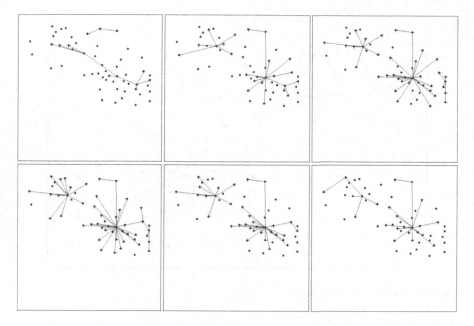

Fig. 8.27 Causation dynamics for the Dengue fever epidemics, steps 1–6

Table 8.13 TDM prediction scores, steps 1–6

TDM Algorithm	
Working hypothesis	378
Right connections	183
Missing connections	3
Sensitivity	98.39 %
Specificity	100.00 %
Global accuracy	99.21 %

We are now in the position to consider the cumulative graph that overlaps the causation effects emerging at the various steps of the process (Fig. 8.27). We can also evaluate the dynamic networks representation by TDM in comparison with the real data. From this comparison, we see that only three links have been missed by TDM. The results are shown in Table 8.13 below and deliver quite interesting implications. The most apparent feature is the level of performance on all of the three dimensions of evaluation. Scores never get below 0.98, and for specificity we even obtain a 100 % level. As expected when discussing the results of the prediction task for the toy problem, it turns out that, when dealing with a real-world problem with a rich and consistent inherent structural grammar, TDM becomes able to learn

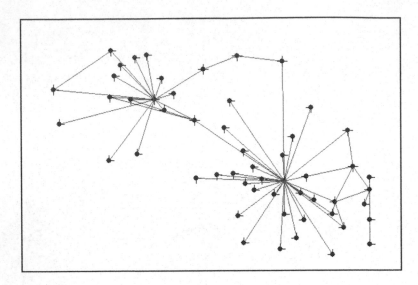

Fig. 8.28 Influence graph, steps 1–6

it very accurately – that is to say, unlike what happens in most test or benchmark evaluations, the performance of the method is *logically* bound to be better in real-world contexts rather than in artificial contexts (Fig. 8.28).

8.9   TWT Applied to the Dengue Fever Epidemics Data

We can now take a further step forward by carrying out the TWT analysis as well. In Figs. 8.29 and 8.30, we report the TWT map vs. the spatial distribution of susceptibility for the first two steps of the process only, comparing it with the distribution of susceptibility at the final step of the process according to the analysis of Ferreira (2010). It is apparent that, already at the first step of the process, the TWT map provides a quite accurate picture of how the final configuration of infection hubs will look like. In particular, the main Southeast hub is clearly found out in all of its complex spatial articulations, and is correctly estimated to be the most virulent, as clearly indicated by the grid tectonics. Moreover, the secondary Northwest hub and the minor, 'satellite' Northeast hub are also clearly sorted out, with their decreasing potential of virulence.

The comparison between Figs. 8.29 and 8.30 presents another interesting feature. TWT actually rebuilds, in the first two steps, the global shape of the epidemic diffusion pattern, even beyond the sampled points. In fact, such shape is quite close to the shape of the *political* map of the region (Fig. 8.30). Figure 8.31 shows a Google map of the region. Its Northern boundaries are naturally marked by lakes and rivers, whereas the Southern boundaries are delimited by rivers and by the main highways.

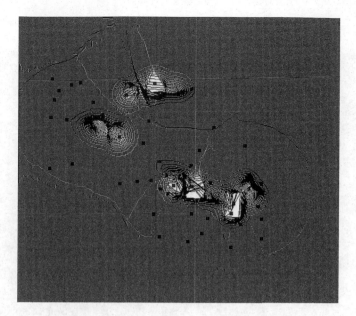

Fig. 8.29 TWT map after the first two steps

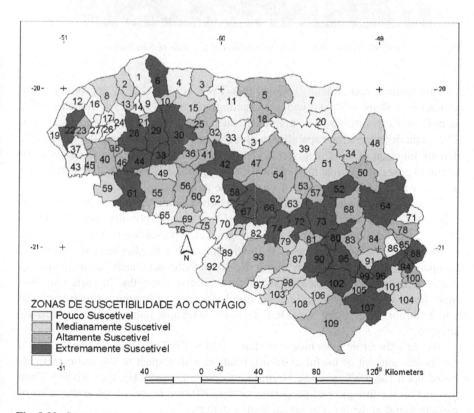

Fig. 8.30 Susceptibility map after step 6

Fig. 8.31 The Google map of the Northwest region of the state of São Paulo

This feature, consequently, shows that TWT is also able to approximate the geometrical shape of a polygon, extracting information from a representative sample of its points. From a mathematical point of view, this could mean that TWT equations have the capability to approximate in a correct way the probability density function of a set of given (hyper-) points. This characteristic can be very useful to generate a global map of the diffusion process, despite the fact that we are generally able to directly monitor only a little part of it.

If we now consider the TWT map obtained from the all of the six steps (Fig. 8.32) and once again compare it with the susceptibility map at step 6 (Fig. 8.30), we find an even more fine-grained representation with the central outbreak-peak block emerging as the main hub with a smaller Eastern secondary component, and with the two minor Northern hubs accurately delineated with neatly differentiated representations of their relative strengths. In particular, this map shows with clarity how the epidemic has been propagating through waves in which minor local hubs appeared as new 'infection pumps' with decreasing strength.

We can therefore conclude that the TDM + TWT analysis seems to add a substantial amount of useful extra information with respect to the static methods based upon the TWC and the Harmonic Center; in particular, they give us deep insights on the fine-grained properties of the diffusion process both at the level of causation and at the level of actual spatial patterns.

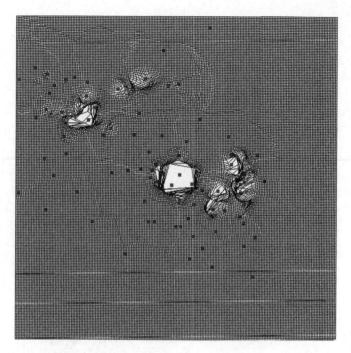

Fig. 8.32 Above: TWT map after step 6

8.10   Neural Networks Step-Wise Prediction

Finally, and again as we did for the toy problem, we now employ a Sine Net to carry out a one-step ahead prediction exercise, i.e., guessing the state of the system at step $n + 1$ given our knowledge of the state at step n. In line of what has happened with the prediction of the link structure for the Dengue fever benchmark as opposed to the toy problem, for the same reasons here we expect again a substantially better performance of the system with respect to the toy problem context. To begin with, we train the network on the first four steps in order to predict the fifth step, following the same methodology that we presented in Sect. 8.4. We obtain the following results (see Table 8.14 and Fig. 8.33):

Results again confirm our intuition: On this very difficult prediction task, the Sine Net delivers substantially better scores on any indicator than in the toy problem context once gain demonstrating that actual structural complexity for this kind of machine learning architecture is a resource rather than a constraint with respect to performance. Notice in particular how, in the present prediction task, precision and accuracy both stand well above 0.99. Similar results emerge from the prediction of step 6 on the basis of the first five steps (see Table 8.15 and Fig. 8.34).

Finally, as a last prediction exercise, we have tried to figure out what the Sine Net would predict for step 7 given the first six steps, i.e., a scenario for which actual data are not available to us. This lack of evidence makes the prediction particularly

Table 8.14 Prediction performance of the Sine Net for step 5 on the basis of the first four steps

Testing	Correct	Total	Errors	%
Sensitivity	31	34	3	91.18
Precision	2,866	2,882	16	99.44
Accuracy	2,897	2,916	19	99.35

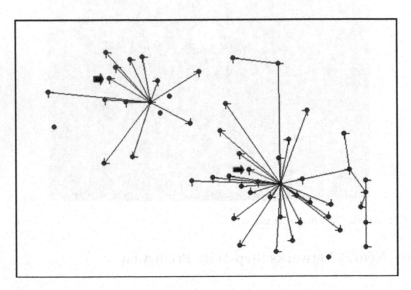

Fig. 8.33 Predicted influence graph at step 5 on the basis of the training on the first four steps. Wrong links are marked in red (color version) with an arrow pointing to it (black/white version)

Table 8.15 Prediction performance of the Sine Net for step 6 on the basis of the first five steps

Testing	Correct	Total	Errors	%
Sensitivity	19	22	3	86.36
Precision	2,881	2,894	13	99.55
Accuracy	2,900	2,916	16	99.45

intriguing, and we look forward to the possibility to acquire more data to make a check on this in future work. Interestingly enough, the Sine Net makes a very sharp and far from obvious prediction for step 7, namely, that the only virulent hub that remains is the main one, whereas the others die out – a circumstance that makes actual checking even more deserving (Fig. 8.35).

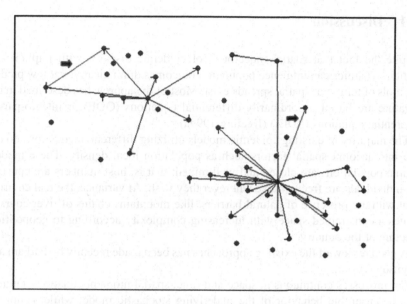

Fig. 8.34 Predicted influence graph at step 6 on the basis of the training on the first five steps. Wrong links are marked in red (color version) with an arrow pointing to it (black/white version)

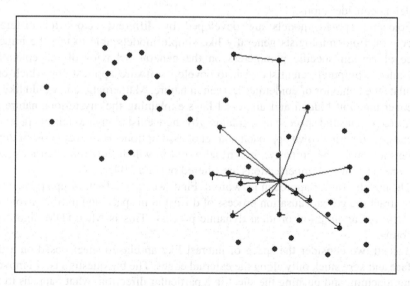

Fig. 8.35 Predicted influence graph at step 7 on the basis of the training on the first six steps. Actual data for step 7 are not currently available

8.11 Discussion

Despite the fact that spatial trajectories of epidemic diseases with implicit and explicit dynamics can influence population outcomes dramatically, yet few predictive tools of temporal spatial spreads exist. Most of the approaches described in the literature are based on ordinarily differential equations (ODE) or integro-partial differential equations (PDEs) (Keeling 1999a).

The majority of existing epidemic models utilizing differential equations do not take into account spatial factors such as population local density. These models assume populations are closed and well mixed; that is, host numbers are constant and individuals are free to move wherever they wish. At variance, the real environment with the presence of natural barriers, like mountains chains of river courses, creates a constrained space with increasing complexity according to geopolitical structure of the country.

A good review of the existing approaches has been made recently by Parham and Ferguson (2005).

For processes continuous in space and time, spatial moment equations summarize the ensemble behavior of the underlying stochastic model while analogous methods have been developed for modeling disease spread on contact networks by capturing how the number of pairs of connected individuals vary over time.

Space may be included in epidemic models within continuous or discrete space frameworks and the choice of model depends on both epidemiological and modeling considerations.

Epidemic spread models are developed by different actors for different objectives. Epidemiologists generally like simple models able to test the importance of certain specific parameters on the general behavior of the epidemic dynamics. Computer scientists wish to develop software applications which can simulate the behavior of epidemics as seen in nature. Mathematicians would like to discover inherent hidden and universal laws explaining the mysterious nature of diffusion in time and space. This explains why there exist many models of epidemic spreads, each with its own approach and set of assumptions. However, most of these models all share one property: the virtual world in which they run is an idealized one where noise and imperfections are filtered out (Fu 2002).

The novelty of our approach is twofold. First, we have a bottom-up approach to reconstruct the global causation process of diffusion in space and in time, avoiding any *a priori* assumption or ideal dynamic process. This is what TDM algorithm performs.

Second, we consider the space of interest like an elastic sheet posed on a flat surface and kept stuck only along the external edges. The big question is: if I move a finger touching and pushing the sheet in a particular direction, what happens to all the points lying on the sheet near or far away to the point below my finger? This is basically what Twisting Theory and the Twisting Algorithm try to solve. Starting from this information a good classifier trained well with the real data can predict future events.

So we now have a variety of techniques with which to study diffusion phenomena from a new perspective, the adoption of a general framework that lends itself quite naturally to applications in a variety of different disciplines, both in the natural and in the social sciences. We feel that more tests need to be carry out to ascertain whether our methodology really works at the levels of accuracy obtained in the present paper in very different disciplinary contexts and for very different problems, and we look forward to tackling this challenge in future research.

It is important to stress, however, that further disciplinary applications of our methodology are of interest not only in order to prove the robustness of the methodology, but possibly also to derive interesting and hopefully cutting-edge results to help us better understand seemingly intractable or ill-defined problems. Among the many possibilities, we are particularly interested in taking a closer look at phenomena such as the diffusion of technology (e.g. Keller 2004) or of innovations (e.g. Walker 1969) and the role of social network structures (for a classical formulation see e.g. Burt 1987), the diffusion of pro-social behavior (e.g. Darkley and Latané 1968), or socio-political attitudes (e.g. Simmons and Elkins 2004), but also at entirely different phenomena such as landslides (e.g. Martin and Church 1997), earthquakes (e.g. Helmstetter and Sornette 2002), and so on. It is an ambitious program, but in our opinion it is an important effort to take a further step along the way toward a truly meta-disciplinary perspective on complex phenomena, their structure, and their implications.

Data Appendix

Table A.1 Matrices of the Strength of Connections from n = 0 to n = 4 in the toy example

Strength of connections between places (Row = Time (n) – Columns= Time(n+1))

n = 0; n = 1	Place 1	Place 2	Place 3	Place 4	Place 5
Place 1	0.000000	0.000000	0.000000	0.000000	0.000000
Place 2	0.000000	0.000000	0.000000	0.000000	0.000000
Place 3	0.000000	0.000000	0.000000	0.000000	0.000000
Place 4	0.000000	0.000000	1.000000	0.000000	0.008928
Place 5	0.000000	0.000000	0.089095	0.016203	0.000000

Strength of connections between places (Row = Time (n) – Columns= Time(n+1))

n = 1; n = 2	Place 1	Place 2	Place 3	Place 4	Place 5
Place 1	0.000000	0.000000	0.000000	0.000000	0.000000
Place 2	0.000000	0.000000	0.000000	0.000000	0.000000
Place 3	0.010040	0.002334	0.000000	1.000000	0.072003
Place 4	0.006473	0.005317	0.000000	0.000000	0.010714
Place 5	0.000994	0.000029	0.000000	0.005165	0.000000

Strength of connections between places (Row = Time (n) – Columns= Time(n+1))

n = 2; n = 3	Place 1	Place 2	Place 3	Place 4	Place 5
Place 1	0.000000	0.040726	0.009736	0.020922	0.010910
Place 2	0.021721	0.000000	0.001697	0.012889	0.000238
Place 3	0.000000	0.000000	0.000000	0.000000	0.000000
Place 4	0.029657	0.034255	1.000000	0.000000	0.058438
Place 5	0.006683	0.000273	0.050911	0.025251	0.000000

Strength of connections between places (Row = Time (n) – Columns= Time(n+1))

n = 3; n = 4	Place 1	Place 2	Place 3	Place 4	Place 5
Place 1	0.000000	0.067575	0.013630	0.031244	0.010560
Place 2	0.016218	0.000000	0.003564	0.028870	0.000345
Place 3	0.004362	0.004752	0.000000	0.948148	0.072407
Place 4	0.010545	0.040599	1.000000	0.000000	0.040402
Place 5	0.004277	0.000582	0.091640	0.048482	0.000000

Tables A.2 Matrices of the plausible connections from n = 0 to n = 4 in the toy example

Presence of connections between places (Row = Time (n) – Columns= Time(n+1))

n = 0; n = 1	Place 1	Place 2	Place 3	Place 4	Place 5
Place 1	0	0	0	0	0
Place 2	0	0	0	0	0
Place 3	0	0	0	0	0
Place 4	0	0	1	0	1
Place 5	0	0	0	1	0

Presence of connections between places (Row = Time (n) – Columns= Time(n+1))

n = 1; n = 2	Place 1	Place 2	Place 3	Place 4	Place 5
Place 1	0	0	0	0	0
Place 2	0	0	0	0	0
Place 3	1	0	0	1	1
Place 4	0	1	0	0	0
Place 5	0	0	0	0	0

Presence of connections between places (Row = Time (n) – Columns= Time(n+1))

n = 2; n = 3	Place 1	Place 2	Place 3	Place 4	Place 5
Place 1	0	1	0	0	0
Place 2	0	0	0	0	0
Place 3	0	0	0	0	0
Place 4	1	0	1	0	1
Place 5	0	0	0	1	0

Presence of connections between places (Row = Time (n) – Columns= Time(n+1))

n = 3; n = 4	Place 1	Place 2	Place 3	Place 4	Place 5
Place 1	0	1	0	0	0
Place 2	1	0	0	0	0
Place 3	0	0	0	1	1
Place 4	0	0	1	0	0
Place 5	0	0	0	0	0

Tables A.3 Tables of the plausible Transformations from n = 0 to n = 4 in the toy example	Trans_0_1			
	Source	TRF	Destination	Strength
	Place_4	->	Place_3	1
	Place_5	->	Place_4	0.016203
	Place_4	->	Place_5	0.008928
	Trans_1_2			
	Source	TRF	Destination	Strength
	Place_3	->	Place_1	0.01004
	Place_4	->	Place_2	0.005317
	Place_3	->	Place_4	1
	Place_3	->	Place_5	0.072003
	Trans_2_3			
	Source	TRF	Destination	Strength
	Place_4	->	Place_1	0.029657
	Place_1	->	Place_2	0.040726
	Place_4	->	Place_3	1
	Place_5	->	Place_4	0.025251
	Place_4	->	Place_5	0.058438
	Trans_3_4			
	Source	TRF	Destination	Strength
	Place_2	->	Place_1	0.016218
	Place_1	->	Place_2	0.067575
	Place_4	->	Place_3	1
	Place_3	->	Place_4	0.948148
	Place_3	->	Place_5	0.072407

References

Becker MH (1970) Factors affecting diffusion of innovations among health professionals. Am J Public Health 60:294–304

Buchanan M (2001) Ubiquity: the science of history, or why the world is simpler than we think. Crown Publishers, New York

Burt RS (1987) Social contagion and innovation: cohesion versus structural equivalence. Am J Sociol 92:1287–1335

Buscema M (2000) Sine Net. A new learning rule for adaptive systems. Semeion technical paper 21, Semeion, Rome

Buscema M, Grossi E (eds) (2009) Artificial adaptive systems in medicine: new theories and models for new applications. Bentham Science Publishers e-book

Buscema M, Grossi E, Breda M, Jefferson T (2009) Outbreaks source. A new mathematical approach to identify their possible location. Physica A 388:4736–4762

Buscema M, Terzi S, Breda M (2006) Using sinusoidal modulated weights to improve feed-forward neural networks performances in classification and functional approximation problems. WSEAS Trans Inf Sci Appl 3:885–893

Darkley JM, Latané B (1968) Bystander intervention in emergencies: diffusion of responsibility. J Pers Soc Psychol 8:377–383

Ferreira MC (2005) Spatial diffusion maps of dengue fever epidemics occurring in southeast of Brazil: a methodology for cartographic modeling in GIS. In: XXIII international cartographic conference, 2005, La Coruña, Espanha

Ferreira MC (2010) Spatial diffusion maps of Dengue fever epidemics occurring in Southeastern Brazil: A methodology for cartographic modeling in GIS. Unpublished, UNICAMP, Campinas

Fu SC (2002) Modelling epidemic spread using cellular automata. http://www.csse.uwa.edu.au/~scfu/caepidemic/

Helmstetter A, Sornette D (2002) Diffusion of epicenters of earthquake aftershocks, Omori's law, and generalized continuous-time random walk models. Phys Rev E 66:061104

Keeling M (1999) Spatial models of interacting populations. In: McGlade J (ed) Advanced ecological theory: principles and applications, Chap 3. Blackwell Science, Oxford, pp 64–99

Keller W (2004) International technology diffusion. J Econ Lit 42:752–782

Martin Y, Church M (1997) Diffusion in landscape development models: on the nature of basic transport relations. Earth Surf Process Landforms 22:273–279

Parham PE, Ferguson NM (2006) Space and contact networks: capturing the locality of disease transmission. J R Soc Interface 3:483–493

Simmons BA, Elkins Z (2004) The globalization of liberalization: policy diffusion in the international political economy. Am Polit Sci Rev 98:171–189

Walker JL (1969) The diffusion of innovation among the American states. Am Polit Sci Rev 63:880–899

Printed in the United States
By Bookmasters